CERAMIC ACOUSTIC DETECTORS

KERAMICHESKIE PRIEMNIKI ZVUKA

КЕРАМИЧЕСКИЕ ПРИЕМНИКИ ЗВУКА

CERAMIC
ACOUSTIC DETECTORS

by

Alevtina Aleksandrovna Anan'eva

Authorized Translation from the Russian

Springer Science+Business Media, LLC

1965

The original Russian text was published by the Academy of Sciences
Press in Moscow in 1963.

Алевтина Александровна Ананьева

Керамические приемники звука

Library of Congress Catalog Card Number 65-11334

ISBN 978-1-4899-4952-3   ISBN 978-1-4899-4950-9 (eBook)
DOI 10.1007/978-1-4899-4950-9

FOREWORD

Piezoceramic, electromagnetic, and electroacoustic transducers are widely used in acoustic technique. Ceramic piezoelectric (piezoceramic) receivers and transmitters are successfully employed in various types of electroacoustic systems, particularly in systems designed for acoustic and underwater acoustic measurements and investigations. The author has been engaged for a number of years in investigating the properties and possible methods of construction of piezoceramic transducers for various purposes, including acoustic and underwater acoustic detectors and transmitters used in measurement systems. The present publication gives the results of our investigations of piezoceramic acoustic receivers. We hope that the book will be of use to workers in the field of acoustic research who are frequently faced with the problem of the design of special acoustic receiving systems for which, in the author's opinion, piezoelectric ceramic is an indispensable material.

The author expresses her sincere gratitude to V. A. Basov, E. V. Vavilov, A. N. Saprygin, and A. V. Sosnov for their assistance in the experiments on which the book is based.

The author also expresses her gratitude to Academician N. N. Andreev and Dr. V. S. Grigor'ev for their sustained interest in the project and for their assistance and valuable advice in carrying out the investigations.

We also consider it our duty to mention the skill displayed by the mechanics A. V. Prakhov and P. D. Kholin which in a number of cases was of decisive importance in the practical realization of the acoustic receivers.

CONTENTS

INTRODUCTION

Piezoelectric materials were first used in acoustic detection systems in 1917 by Langevin who at the end of the first war constructed the first quartz mosaic* [1]. For some time this apparatus, termed the Langevin oscillator, was the only existing piezoelectric device. Other naturally occurring piezoelectric crystals such as tourmaline and zinc blende, the piezoelectric properties of which were already well known, did not find practical application principally because of the high cost of these starting materials. During the first world war, in view of the need for underwater sound receivers for use in echo depth sounders, hydrophones, and similar apparatus, interest developed in the search for an application of synthetic piezoelectric materials, principally Rochelle salt, the piezoelectric properties of which were observed and investigated in 1880 by Pierre and Jaques Curie. Quantitative measurements of the piezoelectric effect in Rochelle salt were first carried out in 1894 by Pockels. Apparently, the first papers proposing the technical application of Rochelle salt in acoustics were published by Nicolson [2]. These papers described the first piezoelectric loudspeakers, acoustic receivers, and microphones. Despite the absence of published information, we can assume that underwater acoustic detectors employing Rochelle salt transducers were also in use at that time. Thus, according to an American source [3] piezoelectric units using Rochelle salt were used in underwater acoustics during 1917-1920 (developed by Western Electric). It is possible that similar work was carried out by the German Atlas-Werke (Bremen).

Up to the time of the second world war the use of Rochelle salt in acoustic devices gradually developed although somewhat slowly. Thus, in 1931 Sawyer [4] described for the first time an improved, technically adequate broadcasting microphone and a loudspeaker employing Rochelle salt transducers. The microphone contained a bimorph piezoelectric transducer, and it should be noted that a piezoelectric transducer in this form was proposed in the USSR by N. N. Andreev as early as 1930 [5].

From this time onwards, the number of published papers dealing with Rochelle salt transducers increased considerably [6-16, etc.], and as early as 1940 the piezoelectric microphone was already proving a serious competitor to the previously employed carbon and electrodynamic broadcasting microphones. However, Rochelle salt possessed a number of disadvantages (hydroscopic nature, low mechanical strength, low Curie point) and this led to the search for other synthetic piezoelectric materials. The work of Valashek, Shul'vas-Sorokin, I. V. Kurchatov, Mason, Fowler, Mueller, Busch, Scherrer, etc., clearly showed that such materials must be sought among substances similar to Rochelle salt in which the piezoelectric coefficients possessed anomalously high values, a fact of great importance for practical utilization in acoustics. The electrical anomalies of Rochelle salt were also of great scientific interest. The unique properties of Rochelle salt and similar substances led to the establishment of the term "seignette electric" (or "ferroelectric") to describe this class of materials.

The search for new ferroelectrics led to the work of Kurchatov, Bloomenthal, and Evans on crystals isomorphic with Rochelle salt and mixed tartrates and to that of Busch and Scherrer (1935) on the ferroelectric properties of KH_2PO_4, the piezoelectric properties of which had been observed by Elings and Terpstra in 1928. The later work of Busch (1938) on the growth and piezoelectric properties of phosphates and arsenates of potassium and ammonium is also of interest.

However, during this period not one of these new piezoelectric materials found practical use in acoustic devices if we except the application of KH_2PO_4 crystals in frequency stabilization.

---

* The piezoelectric properties of quartz were first discovered by the Curie brothers in 1880.

The search for new piezoelectric materials received a sharp stimulus during the second world war due to the shortage of quartz arising from its wide application in underwater acoustic apparatus and in military radio techniques (frequency stabilization, electromechanical filters, etc.). After the war, commencing in 1946, papers began to be published on these developments. We cannot cover the entire literature relating to the investigation of new synthetic crystals and we shall consider here only investigations concerning crystals which found applications in acoustics (e.g., [17-23]). Thus, during the second world war, piezoelectric crystals of ammonium dihydrogen phosphate (ADP) were used in electromechanical filters. This work led to the establishment of a broad group of synthetic piezoelectric materials consisting of ferroelectric materials characterized by high piezoelectric and dielectric constants related to the existence of spontaneous polarization in these materials and also of materials not possessing ferroelectric properties but possessing sufficiently high piezoelectric constants usually accompanied by comparatively low dielectric constants.

Of all these new piezoelectric materials only monoammonium phosphate (ADP), lithium sulfate, and potassium dihydrogen phosphate (KDP) have so far found application in acoustics in the form of mosaics used in underwater transmitter-receivers, i.e., have been used similarly to Rochelle salt. It seems that all the above-mentioned materials suffer from the same disadvantage of low mechanical strength.

The discovery of such a considerable number of synthetic piezoelectric materials led to increased interest in the use of various types of special piezoelectric acoustic receivers and transmitters in acoustic measuring technique. References [24-28] list a number of papers dealing with piezoelectric acoustic receivers. New piezoelectric materials have not yet found proper application in the design of broadcasting microphones[29].

A new stage in the development of piezoelectric acoustic techniques was opened up by the discovery by B. M. Vul (Corresponding Member of the Academy of Sciences of the USSR) and his co-workers of the ferroelectric properties of ceramic barium titanate and the subsequent discovery that polarized ceramic barium titanate possesses piezoelectric properties [30-43]. This discovery marks the commencement of a series of investigations concerning the synthesis of various ceramic piezoelectric materials and their practical application as electromechanical transducers in acoustics and radio technology. Subsequent papers dealt with methods of growing single crystals of barium titanate and other similar materials and showed that such single crystals possessed piezoelectric properties even without artificial polarization. However, up to the present time these single crystals have not found practical application in acoustic techniques.

Generally speaking the piezoelectric properties of polarized ceramic forms of barium titanate and similar materials are not as good as those of synthetic crystals, e.g., Rochelle salt and the phosphates of potassium and ammonium. Consequently, it is necessary to discuss the special properties of piezoelectric ceramics to which they owe their wide and continually increasing application in acoustics. The most important of these properties are: (1) the possibility of producing piezoelectric elements in practically any form by simple technological means; (2) the possibility of creating the required direction of the polarization axis in correspondence with the form of the element by appropriate location of the electrodes; (3) wide choice in methods of siting the electrodes owing to the high dielectric constant. Finally, these materials offer important advantages such as high mechanical strength, low hygroscopic properties, and high dielectric strength.

In the Soviet Union, piezoelectric acoustic receivers utilizing barium titanate ceramic were first developed by Anan'eva and Tsarev [44] working in the acoustic laboratory of the FIAN (Institute of Physics, Academy of Sciences). This receiver was developed in connection with the general problem of constructing small nondirectional sound receivers for measurement purposes. This work served as the starting point for a systematic investigation of the use of piezoceramics in different types of measuring and other sound receivers and transmitters carried out at the acoustic laboratory of the FIAN and at the acoustic Institute of the Academy of Sciences, USSR during the period from 1950 to 1960.

The first foreign papers relating to the use of barium titanate ceramic for acoustic purposes appeared in 1948 [45, 46]. These were followed by a number of papers describing actual sound receivers and transmitters using ceramic transducers. We mention here two papers concerned with high-frequency measurement sound receivers [47, 48] and papers by Koren [49], Mason [50, 51], Johnston [52], Bradfield [53], and other authors [54-56] relating to the use of ceramics in electromechanical transducers, including sound transmitters.

We should not be misled by the comparatively few foreign papers relating to the given problem into believing that interest was lacking in the use of barium titanate ceramic and other similar materials in electroacoustics. This confirmed by the large number of published papers dealing with the piezoelectric properties of ceramic materials [57-74]. Examination of these papers reveals the continually increasing effect of the demands of acoustic technique on chemical and chemical processing developments in the given region. In the Soviet Union, starting in 1950 a large number of investigations were carried out and papers published on new ceramic piezoelectric materials. Of these it is sufficient to mention those by Smolenskii and his co-workers [75-88], Roi [89, 90], Bokov [91, 92], Myl'nikova [93], Insupov and Kosyakov [94, 95], Pasynkov and Vinogradov [96]. A number of research papers concerning the development of piezoelectric ceramic materials and the technology of the preparation of such materials and of piezoelectric elements will be discussed in what follows but only to the extent of direct interest in connection with the design and construction of acoustic receivers. Within the framework of the present publication it would be impossible and scarcely necessary to analyze all the existing voluminous material since improvements in the parameters of piezoelectric ceramics is not important in regard to the solution of pure design problems arising out of the construction of electroacoustic devices.

The present publication is chiefly concerned with work carried out by the author on the design of acoustic receivers of various types using ceramic barium titanate transducers.

The principal material given in Chapters III to VI is prefaced by an introductory section (Chapters I and II) which deals with certain properties of ceramic piezoelectric materials based on barium titanate and with the method of measurement used during investigation of ceramic piezoelectric elements and acoustic receivers employing ceramic barium titanate transducers.

CHAPTER I

DIELECTRIC AND PIEZOELECTRIC PROPERTIES
OF BARIUM TITANATE CERAMIC

Up to the present time barium titanate has only been used in electroacoustic transducers in ceramic form. Barium titanate ceramic is a polycrystalline substance consisting of randomly arranged microcrystals. By subjecting polycrystalline barium titanate to the effect of an external constant electric field it is possible to change the direction of spontaneous polarization in the individual microcrystals to conform to the direction of the external field. After comparatively lengthy exposure to the effect of the external field, the residual polarization is retained when the field is removed with the result that the ceramic behaves like a single-domain single crystal and in particular possesses piezoelectric properties. The process described above is termed "polarizing the ceramic" and a ceramic subjected to the process is termed "polarized."

Nonpolarized ceramic barium titanate subjected to an alternating electric field displays only electrostriction properties and the corresponding ponderomotive effects in this case are quadratic. Polarizing the ceramic results in considerable linearization of the phenomena so that in fact only polarized barium titanate is used in the construction of piezoelectric receivers employing barium titanate ceramic transducers. Since the piezoelectric properties of polarized barium titanate ceramic are finally dependent on the properties of the single crystals of which it is composed, we shall examine the properties of single-crystal barium titanate which are of major interest in the present case.

Single-crystal barium titanate is a ferroelectric since it possesses the property of spontaneous polarization in a given temperature range (below 123°C). X-ray analysis shows that in the temperature range from 4 to 123°C the barium titanate lattice possesses a tetragonal structure. The table of piezoelectric strain coefficients for single crystal barium titanate is characterized by five piezoelectric coefficients which differ from zero.

In this case $d_{31} = d_{32}$ and $d_{24} = d_{15}$. In accordance with this table the equations of the dependence of the piezoelectric effect in a single crystal on strain take the form

$$P_1 = d_{15}Z_x, \qquad P_2 = d_{24}Y_z,$$
$$P_3 = d_{33}Z_z - d_{31}X_x - d_{32}Y_y,$$

where d_{ik} ($i = 1, 2, 3$) are the piezoelectric strain coefficients representing coefficients of proportionality between the components of the piezoelectric polarization P and the components of the mechanical stress producing this polarization.

	X_x	Y_y	Z_z	Y_z	Z_x	X_y
P_1	0	0	0	0	d_{15}	0
P_2	0	0	0	d_{24}	0	0
P_3	$-d_{31}$	$-d_{32}$	d_{33}	0	0	0

TABLE 1

Crystal	Upper Curie point	Crystallographic cut and mode of vibration	Dielectric constant (At room temperature)	Effective piezoelectric strain coefficients (At room temperature)	Dielectric constant at the upper Curie point	Authors
Barium titanate	90		$\varepsilon_1 = 1.5 \cdot 10^3$; $\varepsilon_3 = 1.2 \cdot 10^3$		$\varepsilon_1 = 6.5 \cdot 10^3$; $\varepsilon_3 = 1.9 \cdot 10^3$	Matthias, Hippel [101]
	~117		$\varepsilon_1 = 170 \cdot 10^3$; $\varepsilon_3 = 120 \cdot 10^3$		$\varepsilon_1 = 290 \cdot 10^3$; $\varepsilon_3 = 540 \cdot 10^3$	Matthias
				$d_{33} = 9.5 \cdot 10^{-6}$ $d_{31} = 3.1 \cdot 10^{-6}$		Bond, Mason, McShimin [115]
			$\varepsilon_3 = 200$			Merz [103]
Rochelle salt	24	L-cut, longitudinal thickness	$\varepsilon_L = 170$	$d_L = 1.97 \cdot 10^{-5}$		Mason [22]
		$XY_{t45°}$, longitudinal lengthwise	$\varepsilon_{11}^{T} = 480$	$d'_{12} = 57.5 \cdot 10^{-7}$		
		$YX_{t45°}$, longitudinal lengthwise	$\varepsilon_{22}^{T} = 12$	$d'_{21} = 80 \cdot 10^{-7}$		
Ethylene-diamine	120	YX, longitudinal lengthwise	$\varepsilon_{22}^{T} = 8.2$	$d_{21} = 30.3 \cdot 10^{-8}$		Mason [22]
		ZX, lengthwise shear	$\varepsilon_{33} = 6$	$d_{36} = 50 \cdot 10^{-8}$		

The first figure in the subscript represents the index of the polarization component, the second figure that of the corresponding strain component. The spontaneous polarization in barium titanate is directed along the z axis corresponding to the third index.

Above the temperature of the upper phase transition (i.e., above 123°C), barium titanate does not exhibit spontaneous polarization and consequently is not then a ferroelectric. Megaw [97-100] carried out an x-ray investigation of barium titanate single crystals over a wide range of temperatures and discovered a change in the relationship between the lengths of the a, b, and c axes of the barium titanate lattice with change in temperature. According to the data of Megaw the $BaTiO_3$ lattice transforms at 123°C from the tetragonal structure for which a = b ≠ c to the cubic structure (a = b = c). At 4°C a second phase transition occurs whereby the lattice transforms from the tetragonal to the orthorhombic structure (a ≠ b ≠ c).

The temperatures of the phase transitions are marked by sudden changes in thermal capacity and an increase in dielectric constant. These phenomena are sharply pronounced at the high-temperature phase transition and are observed both in single crystals and in polycrystalline barium titanate. In single-crystal barium titanate the dielectric constant in the direction of the polar axis (ε_3) differs from the corresponding value perpendicular to the polar axis (ε_1).

A number of papers, [101-114] among others, deal with the dielectric properties of single-crystal barium titanate and give the result of dielectric constant and dielectric loss measurements over a wide range of temperature and field intensity values. The piezoelectric properties of single-crystal barium titanate are given in [115], which gives the following values for the piezoelectric strain coefficients:

$$d_{33} = 9.5 \cdot 10^{-6} \text{ CGSE} \quad \text{and} \quad d_{31} = -3.1 \cdot 10^{-6} \text{ CGSE}.$$

Experimentally obtained parameters of single-crystal barium titanate are briefly summarized in Table 1. The considerable scatter in the values is probably due to the fact, mentioned by most of the authors, that the chemical purity of the single crystals was insufficient and in some cases was not even known with certainty.

It is interesting to compare the properties of single-crystal barium titanate with the corresponding properties of other single-crystal ferroelectric materials used in the construction of piezoelectric elements. For this purpose Table 1 gives the parameters of two typical ferroelectrics, i.e., Rochelle salt and ethylenediamine tartrate.

The data given in the table show that the parameters of single-crystal barium titanate are by no means superior to those of the other ferroelectric piezoelectric materials. Its maximum piezoelectric strain coefficient is smaller by approximately one order than the corresponding values for Rochelle salt and ethylene—diamine tartrate whereas the dielectric constant of barium titanate is approximately one order larger than for the other materials all of which is by no means always advantageous. Thus, assuming the future availability of large single-crystal blocks, single-crystal barium titanate would only find application in transducers by virtue of other advantages such as the high temperature of the second phase transition, high mechanical and electrical strength, low hygroscopic properties, etc. However, these advantages are only of value in the case of transducer-driving units, e.g., for acoustic transmitters.

As will be shown later, the dielectric and piezoelectric properties of barium titanate ceramic are not very different from those of single-crystal barium titanate. Thus, the advantage of ceramic piezoelectric materials and, in particular, of barium titanate ceramic in electroacoustics does not depend on the parameters of the materials but is chiefly due to the special technological and constructional possibilities by the use of these materials.

The principal advantage of barium titanate ceramic is the possibility of artificial polarization. Polarized barium titanate ceramic possesses a single symmetry axis of infinite order, the direction of which coincides with that of the polarizing field. If the vector of the polarizing field is directed along the z axis and we cut from the polarized barium titanate a parallelepiped one side of which coincides with the z axis and with arbitrary orientation of the x and y axes, we obtain similitude to a piezoelectric element with tetragonal structure. The corresponding values of the piezoelectric strain coefficient d_{33} and of the coefficients d_{31} and d_{32}

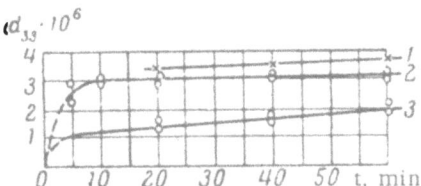

Fig. 1. Dependence of the piezoelectric strain coefficient on polarization time at room temperature. Polarizing field intensity in kV/cm: 1) 25; 2) 15; 3) 7.5.

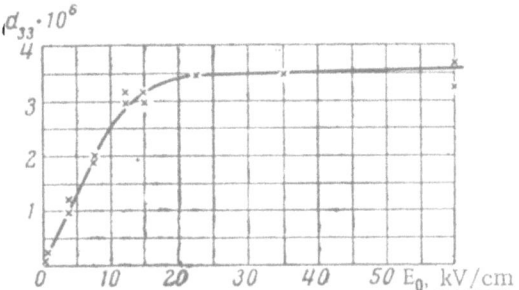

Fig. 2. Dependence of the piezoelectric strain coefficient on the intensity of the constant electric field. Polarization at room temperature.

(along the arbitrarily chosen orthogonal x and y axes) differ only slightly from the corresponding values for single-crystal barium titanate. However, the existence of a symmetry axis of infinite order enables the polarized ceramic to be used in a completely different manner compared with a single crystal.

Another advantage of ceramics is the ease with which by the aid of simple technological means (grinding, pressing, casting) any required form of the piezoelectric element can be obtained, e.g., plane slabs, spherical or cylindrical shells, etc. By polarizing the ceramic in various directions, e.g., longitudinal, transverse, radial, etc., and using different directions of polarization in different regions of the same piezoelectric element it is possible to obtain practically endless variety in the forms of the ceramic elements.

The effect of remanent polarization in a polycrystalline ceramic is similar to the well-known effect of the magnetization of a ferromagnetic by an external magnetic field. Thus, in the foreign literature the seignette-electric properties of barium titanate are termed ferroelectric properties.

The first information concerning the piezoelectric properties of barium titanate ceramic was given in the papers of Adler [37] and Roberts [38]. However, the first quantitative measurements of the piezoelectric effect in a polarized ceramic were made by Rzhanov [40-43].

Ceramics can be polarized under various conditions, e.g., at different values of the intensities of the polarizing field, temperature, polarization time, all of which exert important effects on the final result, i.e., the piezoelectric strain coefficient. At first all piezoelectric ceramics were polarized at room temperature.

We investigated the dependence of the piezoelectric strain coefficient in the direction of the polarizing field (d_{33}) on polarization time and polarizing field intensity at room temperature. The piezoelectric strain coefficients were measured under static conditions by determining the emf at the specimen faces by means of a string electrometer as a function of the force applied in the direction of polarization. The value of the specimen capacity required for calculating the piezoelectric strain coefficient was measured by the dc method of comparison with a calibrated condenser and with alternating current using a bridge or a Q-meter.

Figure 1 shows the effect of the time of polarization at room temperature on the value of the piezoelectric strain coefficient d_{33} for $BaTiO_3$-ceramic specimens prepared in the FIAN dielectric laboratory. The figure shows that polarization for 50 to 60 min is sufficient to almost reach the limit value of the piezoelectric strain coefficient in the given polarization field.

For the same specimens Fig. 2 shows the dependence of the piezoelectric strain coefficient d_{33} on the intensity E_0 of the polarizing field (for polarization for one hour). The limit value of d_{33} obtained for the given specimens was $3.6 \cdot 10^{-6}$ cgs units. A constant electric field with an intensity of 15 to 20 kV/cm is sufficient to obtain the limit value of the coefficient with polarization at room temperature for one hour.

Experiments carried out to determine the electrical breakdown strength showed that the field intensity necessary to achieve breakdown through the specimen varied within wide limits depending on the degree of purity of the starting materials and the exactness of the process of ceramic preparation. The danger of breakdown became serious at a field intensity of approximately 30 kV/cm, although in some specimens breakdown

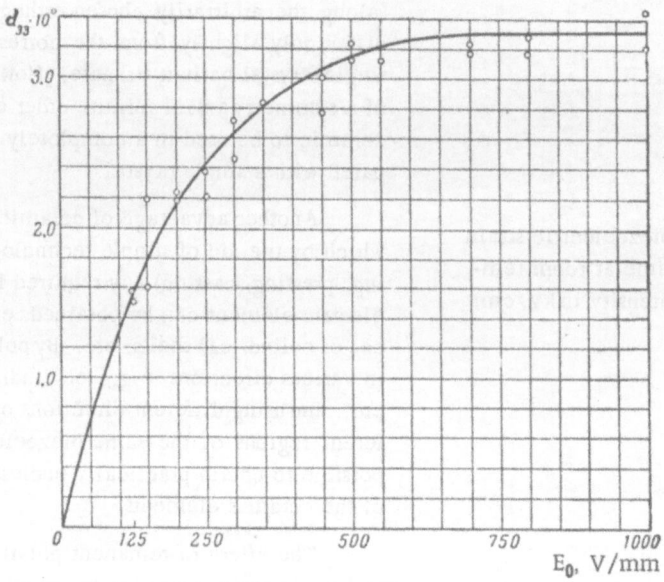

Fig. 3. Dependence of the strain coefficient d_{33} for barium titanate ceramic on the polarizing field intensity with polarization at 123°C.

occurred at field intensities of 15 to 20 kV/cm whereas in carefully prepared specimens (prepared in the Physics Institute of the Academy of Sciences of the USSR) breakdown only occurred at field intensities of 40 to 50 kV/cm [116]. Thus, polarization at room temperature led to unacceptable loss of a large number of specimens.

The required field intensity can be reduced considerably by polarizing at higher temperatures [117-118]. It is of interest that the breakdown intensity for $BaTiO_3$-ceramic was practically independent of temperature over the temperature range of present interest.

Figure 3 shows the value of the piezoelectric strain coefficient in the direction of polarization d_{33} as a function of the polarizing field intensity for polarization at the temperature of the upper Curie point (123°C) and subsequent cooling of the ceramic to room temperature in the polarizing field. The total polarization time including the periods of heating and cooling was not less than 3 h. During polarization the specimen was heated in an oil bath.

Thus, polarization of the ceramic at a higher temperature leads to a considerable reduction (to 6 kV/cm instead of 15 to 20 kV/cm) in the polarizing field intensity required to achieve maximum ordering of the structure, i.e., to obtain the maximum value of the piezoelectric strain coefficient.

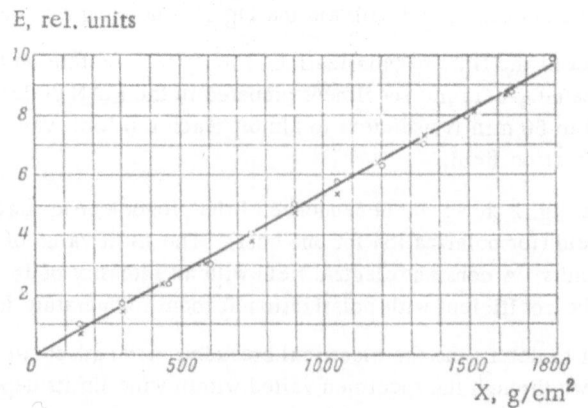

Fig. 4. Dependence of the emf of a polarized barium titanate specimen on the mechanical loading of the specimen.

TABLE 2

Specimen No.	d_{33}, ×10^6	$d_{31} = d_{32}$, ×10^6	$d_{hydr.\ press.}$ $= (d_{33} + 2d_{31})$, ×10^6	Direct measurement of $d_{hydr.\ press.}$, ×10^6	d_{33}/d_{32}
1	2.32	−0.98	0.44	0.47	2.38
2	2.32	−0.98	0.44	0.54	2.38
3	2.18	−0.90	0.38	0.54	2.42
4	3.40	−1.65	0.10	−	2.10

The dependence of the piezoelectric properties on polarization time and the intensity of the constant polarizing field reported by us in [119] were subsequently confirmed by other investigators.

It should be noted that polarization at a higher temperature is not the only means of reducing the polarizing field intensity. It has been shown [120] that the same result can be achieved by cooling to a temperature close to the lower Curie point (for barium titanate ceramic from −4 to +4°C), applying the polarizing field, and heating the specimen to room temperature. However, this method of polarizing ceramics has so far not been widely used in technological practice.

The piezoelectric properties of ceramic BaTiO$_3$ subjected to high-voltage polarization at low mechanical loading are approximately linear. In order to illustrate this fact, Fig. 4 shows the dependence of the emf at the specimen surface on mechanical stress within the limits of 0 to 1800 g/cm^2. At high mechanical loads the linear dependence between the mechanical stress at the surface and the electrical charges is destroyed. According to the data of Vul and his co-workers, the linear dependence between the electrical charges and the mechanical stress in polarized ceramic barium titanate is maintained up to mechanical loads not higher than 100 kg/cm [121, 122]. The same conclusion was reached in [123]. Such high loads are only encountered in

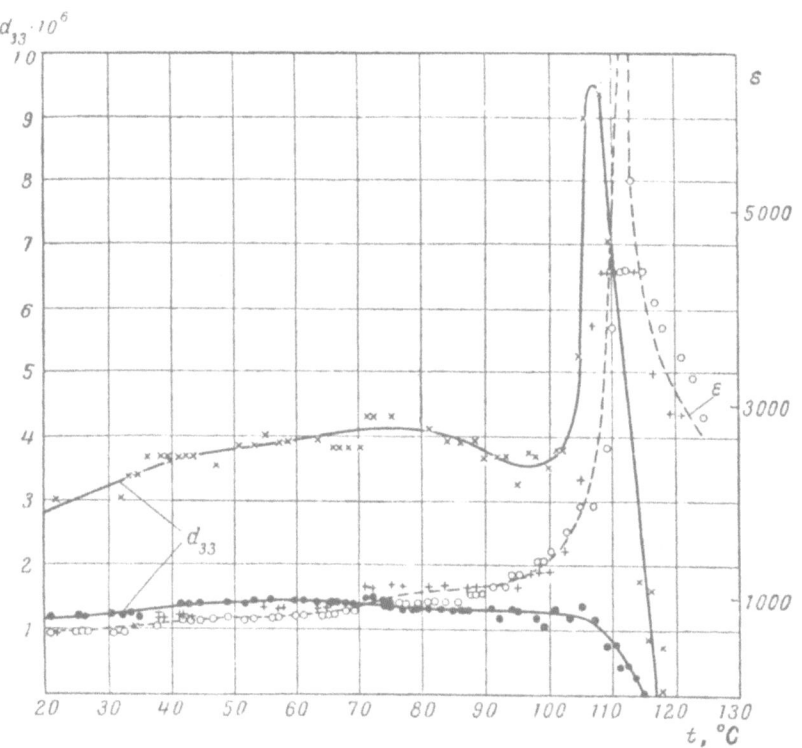

Fig. 5. Temperature dependences of the piezoelectric strain coefficient d_{33} and dielectric constant ε of barium titanate.

TABLE 3

Specimen No.	d_{33}, $\times 10^6$	$d_{32} = d_{31}$, $\times 10^6$	$d_{\text{hydr. press.}} = (d_{33} + 2d_{32})$, $\times 10^6$	d_{33}/d_{32}
5	3.35	−1.39	0.57	2.41
6	3.57	−1.52	0.53	2.35
7	3.52	−1.38	0.76	2.54
8	3.83	−1.51	0.81	2.54
9	3.70	−1.41	0.88	2.59
Mean value	3.6	−1.5	0.7	2.5

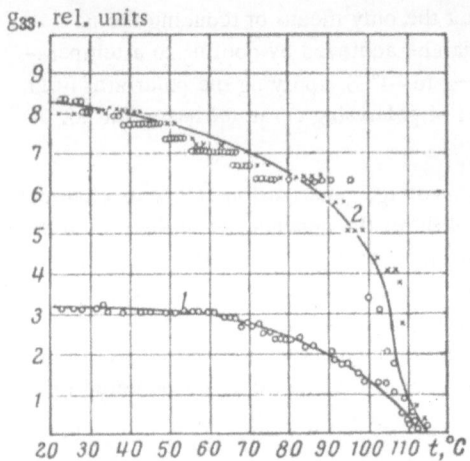

Fig. 6. Temperature dependence of the constant g_{33} for barium titanate. 1) Ceramic polarized at 5 kV/cm; 2) ceramic polarized at 15 kV/cm.

practical acoustic instrumentation in a few exceptional cases. Generally speaking, even in transducer-driver units used in sound transmission the loading is very small and we can assume that the piezoelectric effect in polarized ceramic barium titanate is linear.

In addition to the piezoelectric strain coefficient d_{33}, we also measured the static transverse piezoelectric strain coefficient $d_{31} = d_{32}$ at room temperature. The corresponding data for polarization at room temperature are given in Table 2 which also includes the results of measurements of the strain coefficient for hydrostatic pressure which were carried out with a specimen of cubic form immersed in oil.

Table 3 gives the measured values of the parameters of a ceramic polarized in a field with an intensity $E_0 = 600$ V/cm at +123°C followed by cooling in the high-voltage field. The mean value of the dielectric constant $\varepsilon = 1200$.

The literature frequently gives measurement data for the piezoelectric strain coefficients d_{33} or $d_{31} = d_{32}$, but only in a few instances are both values given for the same specimen. In order to estimate the properties of a ceramic when only one of these strain coefficients is known, the other coefficient is often estimated on the basis of the known experimentally determined ratio of the two coefficients d_{33}/d_{32}. This ratio is given in Tables 2 and 3 for our measurements. Rzhanov's measurements gave a value $d_{33}/d_{32} = 2.28$ while Mason obtained $d_{33}/d_{32} = 2.7$. Kozlabaev [124] arrived at similar results, but the measurements of this author showed that in some ceramic specimens the ratio d_{33}/d_{32} is higher than 3. Berlincourt and Krueger [69] showed that this ratio was dependent to a considerable extent on the porosity of the barium titanate ceramic. At low porosity the ratio decreased to 2.2, increasing to 2.5 with increase in porosity. The ratio d_{33}/d_{31} for barium titanate ceramic and the corresponding ratios for solid solutions of barium titanate with other materials forming isomorphous compounds differ in value. For example, for a compound consisting of 88% $BaTiO_3$ + 12% $CaTiO_3$ the ratio is 3.1. Thus, the discrepancy between our data and the values obtained by Mason, Rzhanov, and Kazlabaev can be explained on the basis of the different quality of the ceramic specimens. Theoretical calculation for single-crystal barium titanate gives $d_{33}/d_{31} = 2$.

It should be mentioned that not only the ratio d_{33}/d_{31} but in general the piezoelectric, dielectric, and elastic properties of ceramics depend to a considerable degree on the quality of the fired ceramic body. Consequently, they depend on the purity of the starting materials (titanium dioxide and barium carbonate) and on the correct method of specimen preparation, i.e., on correct grinding and suitable values of pressure, pressing, and annealing temperature. Our measurements showed that the parameters of industrial ceramic barium titanate (i.e., $BaTiO_3$ without specially introduced impurities) prepared by different organizations differed to some extent. The same discrepancies are encountered in the literature data.

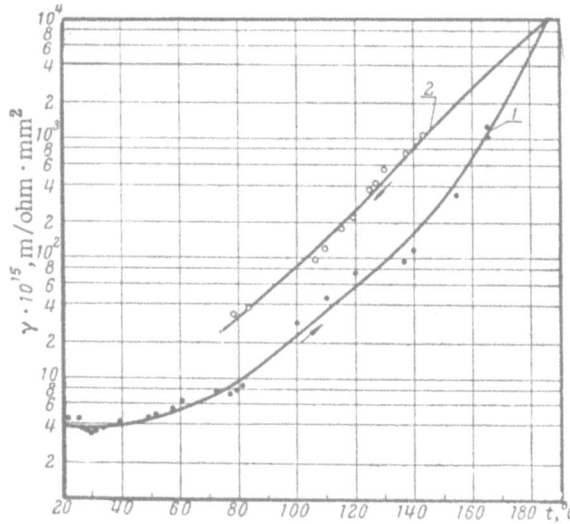

Fig. 7. Temperature dependence of the conductivity of barium titanate ceramic. 1) Measured during heating of the specimen; 2) measured during cooling of the specimen.

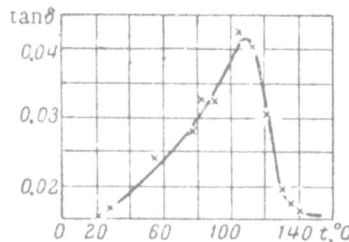

Fig. 8. Temperature dependence of the dielectric losses in barium titanate at a frequency of 5 kc.

In the case of polarized barium titanate ceramic, as for any other piezoelectric material, considerable interest attaches to the dependence of the piezoelectric strain coefficients and dielectric constant on temperature.

Figure 5 shows the temperature dependence of the piezoelectric strain coefficient d_{33} of a polarized ceramic. The lower complex curve relates to a ceramic polarized at room temperature with a polarizing field intensity of 5 kV/cm; the upper curve corresponds to a ceramic polarized to give the maximum value of d_{33} at room temperature ($E_0 = 15$ kV/cm). The broken line curve shows the temperature dependence of the dielectric constant ε.

The sharp increase in the piezoelectric strain coefficient d_{33} in the region of the upper Curie point is of no practical value in the design of acoustic receivers. In fact the sensitivity of a piezoelectric transducer operating as a generator is determined not by the value of the coefficient d_{33} but by the ratio $4\pi d_{33}/\varepsilon$. This ratio is usually designated as the parameter g_{33}.

Figure 6 shows the temperature dependence of the constant g_{33} for barium titanate in relative units. In both cases represented by curves 1 and 2 the polarization was carried out at room temperature. The curves clearly show that g_{33} varies monotonically up to the upper Curie point. In the vicinity of the Curie point the piezoelectric properties of the ceramic disappear totally and irreversibly and can only be renewed by repeated polarization.

Our data for the temperature dependence of the longitudinal piezoelectric strain coefficient d_{33} are in qualitative agreement with data obtained by Rzhanov [42, 43, 125] for the temperature dependence of the transverse piezoelectric strain coefficient d_{31} determined dynamically from the vibrations of a rectangular bar.

In order to judge the applicability of any given piezoelectric ceramic in the field of electromechanical transducers, in addition to the piezoelectric constants we must also know the value and temperature dependences of such parameters as the dielectric constant, conductivity, and dielectric losses.

11

The dielectric properties of barium titanate have been frequently discussed in papers which contain data concerning the dielectric constant and dielectric losses over a wide range of temperature both in single-crystals [101-114] and in polycrystalline $BaTiO_3$-ceramic specimens [30-40, 126-144]. Direct use can be made of these results. Data relating to direct measurement of the conductivity of barium titanate are comparatively rarely encountered in the literature, and for this reason we measured the conductivity of barium titanate ceramic as a function of temperature (Fig. 7). A constant potential difference was applied to the specimen and the current potential difference was applied to the specimen and the current measured by a microammeter.

This figure shows that barium titanate ceramic displays conductivity "hysteresis" typical of imperfect dielectrics. From room temperature up to 50°C the conductivity remains at about $4 \cdot 10^{-15}$ m/ohm $\cdot$ mm^2 and then rises sharply with further increase in temperature. Thus, heating the ceramic to 150°C results in an increase in conductivity of two orders, and with subsequent heating to 185° the increase in conductivity exceeds three orders.

It is interesting to compare this value with the conductivity of such a typical ferroelectric as Rochelle salt. According to Valasek [145] the conductivity of Rochelle salt at room temperature is $(22.0$ to $11.0) \cdot 10^{-14}$ mho/cm (the values are obtained by reversing the sign of the applied emf). Thus, from this point of view barium titanate ceramic is a better dielectric than Rochelle salt.

We also measured the dielectric losses of polarized barium titanate over the temperature range from room temperature to 150°C at frequencies of 50 and 5 Mc. Figure 8 shows the temperature dependence of the loss tangent δ at a frequency of 5 Mc. The upper Curie point is located at 123°C. The measurements were made with a low-intensity alternating field not exceeding 2.5 V/cm.

From the above-mentioned data and also from data given in the literature [43, 111, 129, 136, etc.] we may conclude that the mean value of tan δ for barium titanate ceramic at room temperature varies within limits of 0.01 to 0.03. Ceramics possessing the higher value of tan δ must be considered as being of poor quality.

In the construction of electroacoustic apparatus and especially of measurement sound receivers the constancy of the piezoelectric parameters with time, i.e., the aging properties, are most important. Our experiments and the data of Kozlabaev [124, 146] indicate that barium titanate possesses sufficient stability with time for a number of technical applications. However, Mason [63, 147] shows that for the attainment of a high degree of stability with time piezoceramic units should be artificially aged.

The parameters of piezoelectric barium titanate ceramic are briefly summarized in Table 4 compiled from our own measurements and literature data.

The scatter in the tabulated values is fairly large due probably to the different techniques employed in the preparation of the ceramic specimens by the various investigators. For the purpose of comparison Table 5 gives the parameters of a number of other piezoelectric materials either used in electroacoustics or capable of application in acoustics.

As previously mentioned, the most important parameter with respect to the application of piezoceramics as transducer-generators is the ratio $4\pi d_{33}/\varepsilon = g_{33}$. The mean value of this ratio for barium titanate ceramic may be taken as $3.8 \cdot 10^{-8}$ cgs units. On the other hand, for example in Rochelle salt with an $XY_{t\,45°}$ cut, the parameter g'_{21} which plays the same part is equal to $85 \cdot 10^{-8}$ cgs units. Even in quartz, the corresponding parameter $g_{11} = 19.26 \cdot 10^{-8}$ cgs units. The low value of the parameter g_{33} in barium titanate ceramic is due to the very high value of the dielectric constant. Thus, it is desirable that the ratio d_{33}/ε should be increased either by increasing the piezoelectric strain coefficient or decreasing the dielectric constant. One possible method of reducing the dielectric constant is the introduction of various impurities into the ceramic mix prior to firing. It is not necessary that the impurities fall into the category of ferroelectrics. We investigated the use of aluminum oxide (up to 10 to 15%) as impurity which resulted in a ceramic with good firing properties, with no displacement of the upper Curie point and a considerably reduced dielectric constant. However, this was associated with a simultaneous decrease in the piezoelectric strain coefficient so that the ratio d_{33}/ε remained practically the same as for the barium titanate ceramic [157, 165]. An attempt to obtain a piezoelectric material by grinding ferroelectric barium titanate ceramic after firing and then forming the powder thus obtained with the aid of a binder (e.g., methylmethacrylate) and subsequent polarization after polymerization

TABLE 4

Dielectric constant ε	Piezoelectric constants						Density, g/cm³	Elastic constants	Electro-mechanical coupling	Curie point, °C	Authors	Remarks
	d_{33} ×10⁶	d_{31} ×10⁶	d_{15}	g_{33} ×10⁸	g_{31} ×10⁸	g_{15}						
1600	3.2	−1.4		2.51	−1.1		6.2	$E = 1.15 \cdot 10^{12}$ d/cm²			Rzhanov [40, 42, 125]	
	5.4	−2.25									Jaffe [57]	
	3.65	−1.4	5.35								Vul, Bogdanov, Rozbash [148]	
1420–1565	3.2–4.3						5.5–5.71	Q = 75–164	0.33–0.41	125	Egertone, Koonce [65]	Ceramics of industrial grade materials — The different values correspond to the different firing temperatures
	0.49–4.45		5.00–7.50				5.37–5.72	Q = 80–232	0.34–0.37			Chemically pure ceramic
ε_{33} = 900–1500, ε_{11} = 1500–1800	3.00–6.7	−2.35 −1.1		−8.1 −7.1	2.5–1.2	4.2–6.3	5.5–5.8				Bronnikova, Rozbash [149]	
1600	3.5	−1.3									Roi [89, 90]	
	5.75	−2.35	7.8								Berlincourt [66]	
	5.73	−2.37	8.1								Bechman [67]	
	4.0	−1.5	5.5								Bogdanov, Vul, Timonin [150, 151]	
1800	4.2						5.7				Smolenskii, et al. [152]	
	5.7	−2.34									Berlincourt, Krueger [69]	Industrial ceramic
	6.87	−2.84					5.85					Pure ceramic
2000, ε_{11}^{T} = 1436, ε_{33}^{T} = 1680, ε_{11}^{S} = 1123, ε_{33}^{S} = 1256		−2.25			−1.41				k_{33} = 0.185		Herbert [70]	Industrially pure ceramic
	5.74	−2.37	8.1		−2				k_{33} = 0.494, k_{31} = 0.208, k_{15} = 0.466		Berlincourt and Jaffe [153]	
1280		1.5		4.21	1.47			$E = 11.0 \cdot 10^{11}$ d/cm², Q = 400	k_{33} = 0.52, k_{31} = 0.22, k_p = 0.22	130	Rez, Smazhevskaya, Kachacheva [154]	
1700	5.7	−2.34		4.21	−1.76					115	Crawford [71]	
1200		−1.4			−1.46		5.35				Serova, et al. [155]	Parameters recommended to industry
1700		−2.34			1.76		5.7		k_p = 0.36	113	Berlincourt [166]	
ε_{33} = 1872, ε_{11} = 1596	5.73	−2.35	7.8	3.84	−1.575		5.77		k_{33} = 0.496, k_{15} = 0.46		[156]	
ε^{T} = 1600, ε^{S} = 1200	5.4	−2.25		4.3	−1.8		5.5	$E = 1.1 \cdot 10^{12}$ d/cm², σ = 0.31	k_{33} = 0.45, k_t = 0.31	110	Anan'eva [157]	
ε = 1200		−1.4		3.77			5.3	$s = 1.1 \cdot 10^{-12}$		123		Ceramics made from industrial grade raw materials
$_1$750–5900	3.6	−2,4					5.9	$E = 1.25 \cdot 10^{12}$ d/cm²			Anan'eva et al. [158, 159]	High purity ceramic

ТАБЛ 2

Material	Chemical formula	Limit working temp., °C	Density, g/cm³	Crystallographic cut and vibration mode	Effective piezoelectric strain coefficient ×10⁶	Dielectric constant	Piezoelectric constant ×10⁸	Elastic constant ×10¹²	Authors	Notes
Potassium tartrate (racemate of potash)	$K_2C_4H_4O_6 \cdot \frac{1}{2}H_2O$	80	1.987	$XY_{t45°}$; longitudinal lengthwise	$d'_{12}=d'_{13}$ $=\frac{d_{14}}{2}=34.5$	$\varepsilon^T_{11}=6.49$	$g'_{12}=g_{13}=\frac{g_{14}}{2}=66.3$	$s'_{22}=s'_{33}=4.01$	Mason[22]	
Potassium dihydrogen phosphate KDP	KH_2PO_4	100		ZX; face shear	$d_{36}=69.6$	$\varepsilon^T_{33}=23.75$ $\varepsilon^S_{33}=21.8$	$g_{36}=39.4$	$s_{66}=16.1+16.4$	Mason[17, 22]	Cut seldom applicable
				Longitudinal lengthwise	$d'_{31}=34.8$	$\varepsilon_{33}=21.8$	$g'_{31}=20.2$	$s'_{11}=4.77$		
Ammonium dihydrogen phosphate (mono-ammonium phosphate ADP)	$NH_4H_2PO_4$	100	1.803	$ZX_{t45°}$; longitudinal lengthwise	$d'_{31}=74$	$\varepsilon_{33}=15.7$	$g'_{31}=59.2$	$s'_{11}=5.1$	Mason[17, 22]	
				ZX; face shear	$d_{36}=148$	$\varepsilon_{33}=15.7$	$g_{36}=118.5$	$s_{66}=16.5$		
Triglycine sulfate	$(NH_2CH_2COOH)_3 \cdot H_2SO_4$	49	1.68	Longitudinal lengthwise	For nonpolarized $d_{32}=76$ For polarized $d_{32}=150$	$\varepsilon_{22}=43$	$g_{32}=44.5$	$s_{33}=93.2$	Konstantinova et al.[160, 161]	
Terpin monohydrate	$C_{10}H_6(OH)_2 \cdot H_2O$	116	1.11	ZX; longitudinal thickness	$d_{33}=6.6$	$\varepsilon_{33}=3.2$	$g_{33}=25.9$	$s_{33}=8.57$	Chumakov et al.[164]	
				Longitudinal lengthwise	$d_{32}=10.6$	$\varepsilon_{33}=3.2$	$g_{32}=41.6$	$s_{22}=11.986$		
				Shear thickness	$d_{24}=17.3$	$\varepsilon_{22}=2.8$	$g_{24}=76.2$	$s_{44}=40.9$		Cut not easily applied in practice
Benzophenone	$(C_6H_5)_2CO$	47	1.219	$ZX_{t45°}$; longitudinal lengthwise	$d_{31}=\frac{d_{36}}{2}=30.5$	$\varepsilon_{33}=3.7$	$g'_{31}=50.2$	$s'_{11}=13.03$	Chumakov et al.[163]	
Monohydrate -rhamnose		126	1.417	YZ; longitudinal lengthwise	$d_{33}=13.7$ $d_h=16.4$	$\varepsilon_{22}=2.9$ $\varepsilon_{33}=6$	$g_{33}=58.6$ $g_h=34.3$	$s_{33}=5.86$	Chumakov et al.[164]	Cut used in cases of hydrostatic pressure
Quartz	SiO_2	300	2.65	Curie or XY cut	$d_{11}=6.27$	$\varepsilon_{11}=4.49$	$g=17.52$	$s_{11}=2.649$	Curie, Voigt, Riecke	
					$d_{11}=6.93$	$\varepsilon_{11}=4.58$	$g_{11}=19.26$	$s_{11}=1.277$	Mason[22]	
Tourmaline					$d_{33}=5.8$	$\varepsilon_{33}=6.05$	$g_{33}=15.15$		Riecke, Voigt	
					$d_{33}=5.5$	$\varepsilon_{33}=7.5$	$g_{33}=9.25$	$s_{33}=0.636$	Mason[22]	
Lithium sulfate	$Li_2CO_4 \cdot H_2O$	90	2.052	XY	$d_{22}=49$	$\varepsilon^T_{22}=6.5$	$g_{22}=94.7$	$s_{22}=2.25$	Bechman[23]	
				XY	$d=43.3$	$\varepsilon^T_{22}=6.5$	$g_{22}=6.5$	$s_{22}=2.13$	Mason[22]	
Rochelle salt	$NaKC_4H_4O_6 \cdot 4H_2O$	24	1.771	L-cut; longitudinal thickness	$d_L=197-200$	$\varepsilon^T_L=170$	$g_L=147$	$s_L=5.6$		
				$XY_{t45°}$; longitudinal lengthwise	$d'_{12}=57.5$	$\varepsilon^T_{11}=480$	$g'_{12}=15$	$s'_{11}\approx 6.6$		
				$YX_{t45°}$; longitudinal lengthwise	$d'_{21}=80$	$\varepsilon^T_{22}=12$	$g'_{21}=85$	$s'_{11}\approx 8.2$		
Ethylenediamine tartrate EDT	$C_6H_6N_2O_6$	120	1.538	YX; longitudinal lengthwise	$d_{21}=30.3$	$\varepsilon^T_{22}=8.2$	$g_{21}=46$	$s_{11}=3.34$		
				ZX; face shear	$d_{36}=50$	$\varepsilon_{33}=6$	$g_{36}=1045$	$s_{11}=3.34$		

TABLE 6

Composition	Dielectric constant	Piezoelectric constants					Density, g/cm³	Elastic parameters	Electromechanical coupling coefficient	Curie point, °C	Authors	Remarks
		d_{33}	d_{31} ×10⁶	d_{15}	g_{33}	g_{31} ×10⁸						
(Ba$_{0.9}$Zr$_{0.1}$)TiO$_3$	ε = 1500	1.8			1.5–0.5						Smolenskii, et al, [83]	The values k, ε, and d$_{13}$ are given for t = 20°, the values of ε are approximate values obtained from graphs
(Ba$_{0.98}$Pb$_{0.02}$)TiO$_3$	1000		−2.5			3.14			0.26	180	Myl'nikova [93]	
(Ba$_{0.94}$Pb$_{0.06}$)TiO$_3$	800		−2.4			3.76			0.28			
(Ba$_{0.92}$Pb$_{0.08}$)TiO$_3$	700		−2.1			3.77			0.27			
(Ba$_{0.88}$Pb$_{0.12}$)TiO$_3$	600		−1.3			3.72			0.165			
(Ba$_{0.98}$Ca$_{0.11}$)TiO$_3$	670		−0.95			1.77				130	Rez et al. [154]	
(Ba$_{0.78}$Ca$_{0.13}$Pb$_{0.09}$)TiO$_3$	450		−0.5			1.39				150		
(Ba$_{0.8}$Pb$_{0.2}$)TiO$_3$	220		−0.4			2.29				250		
Pb(Zr$_{0.55}$Ti$_{0.45}$)O$_3$	600		−1.1			−2.29				250		
(Ba$_{0.94}$Ca$_{0.06}$)TiO$_3$					4.4	−4.33					Pasynkov, Vinogradov [96]	
(Ba$_{0.88}$Ca$_{0.08}$Pb$_{0.04}$)TiO$_3$					4.65	−1.33						
PZT-4	ε$_1$ = 1360 ε$_3$ = 1200	7.68	−3.33	13.5	8.05	−3.46	7.5	Q = 600	k$_{33}$ = 0.64 k$_p$ = 0.52 k$_{31}$ = 0.31 k$_{15}$ = 0.65	340	Berlincourt et al. [168]	
PZT-5	ε$_1$ = 1285 ε$_3$ = 1500	9.6	−4.2	14.7	8.05	−3.2	7.5	Q = 75	k$_{33}$ = 0.675 k$_p$ = 0.53 k$_{31}$ = 0.32 k$_{15}$ = 0.655	360		
(Ba$_{0.95}$Ca$_{0.05}$)TiO$_3$ (wt. %)	ε$_1$ = 1280 ε$_3$ = 1200	4.5	−1.74	7.71	4.71	−1.82	5.5	Q = 500 E$_{11}$ = 11.6·10^{11} E$_{22}$ = 11.1·10^{11} d/cm² E$_{33}$ = 4.4·10^{11} d/cm²	k$_{33}$ = 0.49 k$_p$ = 0.325 k$_{31}$ = 0.19 k$_{15}$ = 0.495	115		
(Ba$_{0.8}$Pb$_{0.12}$Ca$_{0.08}$)TiO$_3$	400	1.8	−0.6		5.65	−1.88	5.4	Q = 1200	k$_{33}$ = 0.34 k$_p$ = 0.19 k$_{31}$ = 0.113 k$_{15}$ = 0.3	140		
(Ba$_{0.95}$Ca$_{0.05}$)TiO$_3$ + 0.75CoCO$_3$	1420			1.77			5.7		k$_p$ = 0.31 k$_{31}$ = 0.182	105	Berlincourt et al. [168]	
Lead zirconate-titanate	ε = 1200		−3.93			−1.41		E$_{11}$ = 0.85·10^{12} d/cm²	k$_{31}$ = 0.355 k$_p$ = 0.6	320	Crawford [71]	Composition recommended for high-power transmitters
Lead zirconate-titanate	ε = 1500	9.6	−4.2		8.05	−3.51		E$_{11}$ = 0.675·10^{12} d/cm²	k$_{31}$ = 0.318 k$_p$ = 0.54 k$_{33}$ = 0.675	350		Composition recommended for wide-band receivers
Ba$_{0.9}$Pb$_{0.1}$TiO$_3$	500	5.7	−0.69					Q = 800 E$_1$ = 1.2·10^{12}	k$_{31}$ = 0.12 k$_{33}$ = 0.36 k$_p$ = 0.2	150	Jaffe [169]	
PbTi$_{0.45}$Zr$_{0.55}$O$_3$	500	5.25	−1.68					Q = 300 E$_1$ = 0.75·10^{12}	k$_{31}$ = 0.23 k$_{33}$ = 0.55 k$_p$ = 0.39	350		
PZT-4	1200	3.9	−3.15					Q = 600 E$_1$ = 0.35·10^{12}	k$_{31}$ = 0.29 k$_{33}$ = 0.63 k$_p$ = 0.5	340		

TABLE 7

Composition	Dielectric constants	Piezoelectric constants						Density, g/cm³	Elastic parameters	Electromechanical coupling coefficient	Curie point, °C	Authors
		d_{33}	d_{31} $\times 10^6$	d_{15}	g_{33}	g_{31} $\times 10^8$	g_{15}					
$(Pb_{0.6}Ba_{0.4})Nb_2O_6$	1250		−1.3			−1.31					280	Rez et al. [154]
$PbNb_2O_6$	$\varepsilon = 800$	2.8			4.25						570	Goodman [61]
$Na_{1.6}Cd_{0.2}Nb_2O_6$	$\varepsilon = 500$	5.25	−1.68						$E_1 = 0.75 \cdot 10^{12}$	$k_{31} = 0.23$ $k_{33} = 0.55$ $k_p = 0.39$	350	Jaffe [169]
$PbNb_2O_6$	$\varepsilon = 225$	2.4	−0.33						$E_1 = 0.8 \cdot 10^{12}$	$k_{31} = 0.045$ $k_{33} = 0.42$ $k_p = 0.07$	570	
$Pb_{0.8}Sr_{0.2}Nb_2O_6$	440		−1.3			−3.64				$k_r = 0.26$	450	Isupov, Kosyakov [95]
$Pb_{0.7}Sr_{0.3}Nb_2O_6$	502		−1.6			−4.02				$k_r = 0.31$	380	
$Pb_{0.6}Sr_{0.4}Nb_2O_6$	755		−1.6			−2.64				$k_r = 0.26$	310	
$Pb_{0.55}Sr_{0.45}Nb_2O_6$	1060		−1.5			−1.76				$k_r = 0.22$	250	
$Pb_{0.7}Ba_{0.3}Nb_2O_6$	640		−1.1			−2.14				$k_r = 0.19$	345	
$Pb_{0.6}Ba_{0.4}Nb_2O_6$	1190		−1.7			−1.76				$k_r = 0.27$	290	
$Pb_{0.5}Ba_{0.5}Nb_2O_6$	1205		−1.1			−1.13				$k_r = 0.16$	315	

of the filler [157, 165] resulted in a sharp decrease in the piezoelectric strain coefficient accompanied by a decrease in the parameter g_{33}. According to our data it is possible to achieve values of $d_{33} = 0.02$ to $0.09 \cdot 10^{-6}$ cgs units and of $g_{33} = 0.36$ to $0.72 \cdot 10^{-8}$.

Similar investigations were carried out by Zheludev [166, 167]. According to his data, the maximum mean value of g_{33} obtainable with this type of piezoelectric material is $2.7 \cdot 10^{-8}$ which exceeds our own results by a small factor.

A method which offers better results is the introduction into the ceramic composition of small amounts of other ferroelectrics such as titanates of lead, tin, strontium, etc. This leads to the formation of solid solutions with new quantitative characteristics, e.g., with marked displacement of the Curie point, and for which the values of the piezoelectric strain coefficients and dielectric constants differ from the characteristics of the ferroelectric components forming the given solid solutions.

A great many papers have been published on the dielectric, mechanical, and piezoelectric properties of complex ceramic materials. We have investigated the binary compounds $Ba(TiSn)O_3$, $(BaSr)TiO_3$, and $(BaSr)TiO_3$ containing CoO (0.5%) [119, 157, 165]. Roi investigated the compounds $(BaSr)TiO_3$, $Ba(TiSn)O_3$, and $Ba(TiZr)O_3$ [89, 90]. Smolenskii and his co-workers investigated a wide class of compounds of the type $(BaPb)TiO_3$, $(BaZr)TiO_3$, etc. [75-88]. Mason [63, 147] investigated the binary compounds $(BaPb)TiO_3$ and the ternary compounds $(BaPb)TiO_3 + CaTiO_3$. Berlincourt, Krueger, Crawford, and Jaffe [66, 69, 71] investigated $PbTiO_3 + PbZrO_3$ compounds. A large number of papers deal with ceramic compounds which do not contain titanium, e.g., niobates, tantalates, etc. [61, 74, 94].

Published data concerning the piezoelectric parameters (at room temperature) of complex ceramic compounds based on barium titanate are summarized in Table 6.

Lead zirconate-titanate solid solutions are of considerable interest from the point of increasing the piezoelectric constant g_{33}. Thus, according to the data of [168] a value of $g_{33} = 8.05 \cdot 10^8$ was obtained for a ceramic containing 45% lead titanate and 55% lead zirconate, i.e., a value more than twice as great as the mean value for barium titanate. In this case the value of the dielectric constant was only slightly different from that of barium titanate ceramic. A value of $g_{33} = 4.4 \cdot 10^{-8}$ for a ceramic compound containing 94% $BaTiO_3$ and 6% $CaTiO_3$, i.e., slightly higher than the corresponding value for barium titanate ceramic, was obtained in [96]. Such compounds will no doubt find wide application in electroacoustics.

Although a number of new complex ceramics do not possess higher values of g_{33} in comparison with barium titanate ceramic, they nevertheless possess certain special advantages in connection with the construction of piezoelectric transducers. For example, special interest attaches to compounds with increased coercive force (solid solutions of barium titanate and lead titanate [93]) and compounds in which the temperature dependence of the parameters is less pronounced than in $BaTiO_3$ ceramic with no additives (e.g., 80% $BaTiO_3$ + 12% $PbTiO_3$ + 8% $CaTiO_3$ or 84% $BaTiO_3$ + 8% $PbTiO_3$ + 8% $CaTiO_3$ [63, 147]); solid solutions of the type $(BaCa)TiO_3$ and $(BaPbCa)TiO_3$ give a smooth temperature dependence of the parameters in the neighborhood of the low temperature phase transition [63, 96, 174]. This is a matter of considerable importance from the point of view of acoustic receivers designed to operate at low external temperatures.

Piezoelectric ceramics in which the second phase transition occurs at high temperature, e.g., lead niobates and lead tantalates, etc. [94, 95, 175], are of interest for certain applications. At the present time it is possible to obtain solid solutions with an upper Curie point in the vicinity of 560°C together with satisfactory values of the piezoelectric parameters. However, most compounds with high Curie points possess increased conductivity at high temperatures which leads to difficulty in polarization. This difficulty has been successfully overcome by the addition of bismuth, cadmium, cerium, tungsten, etc.

In recent years a number of papers have dealt with the effects of technological factors (firing temperature, duration of holding at maximum temperature, dispersity of the polycrystalline powder, etc.) on the properties of barium titanate ceramic [65, 176]. Interesting papers have also appeared on the synthesis of barium titanate from pure starting components [65, 158, 177]. These investigations showed that chemically pure barium titanate ceramic possessed better properties with a ceramic prepared from industrially pure starting materials [159].

Despite the fact that all the piezoceramics so far investigated can be prepared in forms dimensioned to permit their use in acoustic receivers, there is no doubt that in the very near future technological difficulties will be overcome and new piezoelectric materials will find practical application in electroacoustics. In this connection Table 7 summarizes the dielectric and piezoelectric properties of a number of new ceramic materials which differ from barium titanate and its solid solutions.

On the basis of existing data concerning the parameters of polarized barium titanate ceramic we shall use the following values: dielectric constant $\varepsilon = 1200$, piezoelectric strain coefficient $d_{33} = 3.6 \cdot 10^{-6}$ cgs units, piezoelectric strain coefficients $d_{31} = d_{32} = -1.4 \cdot 10^{-6}$ cgs units. These values correspond to the piezoceramic principally used in the experimental acoustic receivers. This material was prepared from industrial raw materials by the process described for example in [178-181].

Finally, the method of calculating the piezoelements introduced in the following chapters are applicable to ceramics of any composition; numerical data for the parameters of piezoelements and acoustic receivers constructed from other ceramics can be obtained by substituting in the formulae the values of the piezoelectric, dielectric, and other constants relating to the given piezoceramic.

CHAPTER II

METHODS OF DETERMINING SOUND RECEIVER CHARACTERISTICS

1. The Basic Characteristics of Sound Receivers

Any sound receiver operating as an electroacoustic transducer can usually be described in terms of the following characteristics.

1. Frequency response. In the present case of piezoelectric sound receivers the frequency response extends from zero frequency to a given upper limit frequency so that the response at zero frequency or static response is one of the most important parameters of piezoelectric sound receivers.

2. Frequency dependence of the electrical impedance. In most cases it is necessary to know not only the modulus but also the phase of the impedance as a function of frequency.

3. Directivity characteristics at various selected frequencies. Knowledge of the peak characteristics of sound receivers is only required in special cases, e.g., the measurement of very large sound pressures; generally, measurement and other types of sound receivers operate within the limits of the linear region of the input-output characteristic.

Various methods exist for determining sensitivity as the basic parameter characterizing the sound receiver. In what follows we shall be concerned principally with the field sensitivity determined as the ratio of the emf developed by the sound receiver to the acoustic pressure which would exist in the sound field in the absence of the sound receiver. Once we know the dependences of the field sensitivity and electrical impedance on frequency, we can calculate the actual sensitivity for any condition of electrical loading.

The use of such values as pressure sensitivity as the principal characteristic is less convenient since they are not directly characteristic of the acual sensitivity of the sound receiver and require additional data concerning the frequency dependences of the diffraction correction and mechanical or acoustic impedance of the receiver. In many cases experimental determination of such data is difficult and the work of measurement is no less than in the case of direct measurement of the field sensitivity.

Naturally, the frequency dependence of the field sensitivity must be determined for a given location of the sound emitter with respect to the receiver. If the directivity characteristic of the receiver possesses marked axial symmetry, then it is convenient to locate the sound emitter on the axis of symmetry of the directivity characteristic. In this case we obtain the axial sensitivity. Knowing the frequency dependence of the axial sensitivity we can construct directivity curves normalized with respect to the axial sensitivity, i.e., in relative units.

In the absence of axial symmetry it is still necessary to select a given direction with respect to the receiver to serve as a reference: the frequency dependence of the sensitivity is then determined for this direction. In this case, the directivity curves are normalized with respect to the sensitivity for the reference direction. Thus, in the case of cylindrical sound receivers the direction perpendicular to the axis of the cylinder is selected as the reference direction.

In order to measure the field sensitivity, additional information is required concerning the nature of the sound field used during the measurements. It is generally assumed that the receiver is located in the field of a plane sinusoidal sound wave. This fact must be borne in mind in the practical realization of the corresponding measuring apparatus.

When using sound receivers to measure very small sound pressures, the question arises as to the minimum sound pressure which can be measured with the aid of the given sound receiver and preamplifier. However, as a rule the receiver's self-noise is very small. Thus, Goncharov and Krasil'nikov [182] showed that the inherent noise of piezoelectric elements, including barium titanate ceramic elements, was at least one order of magnitude smaller than the self-noise of the electronic amplifier system connected to the input. Therefore, the minimum measured sound pressure cannot be considered as a characteristic of the sound receiver proper.

Finally, such technical parameters as the maximum permissible external static pressure, mechanical, and vibrational strength, quality of hermetic sealing, etc., are also of interest. Since such parameters are neither acoustic nor electroacoustic in nature, we shall not discuss them here since they can be determined by normal methods of measurement.

2. Measurement of the Static Sensitivity of the Sound Receiver

As stated above, the static sensitivity is the response of the receiver at zero frequency, and its measurement enters logically into the series of measurements made to determine the overall frequency response. Generally, however, the static sensitivity is measured by methods which differ from those used for the measurement at frequencies different from zero. In fact, at zero frequency the diffraction corrections are equal to zero and the field sensitivity coincides with the pressure sensitivity. Therefore, it becomes possible to determine the static sensitivity of a sound receiver simply by placing the receiver in a closed vessel (in which by some means an excess pressure can be created) and measuring the constant emf set up at the transducer output.

The static sensitivity is usually measured at relatively high excess pressures in order to avoid the use of dc amplifiers and enable the output emf to be measured by means of a string or some other type electrometer. However, the excess pressure must not exceed the limit value set by the linear region of the input-output characteristic of the sound receiver. In the case of piezoceramic elements nonlinearity appears only at very high mechanical stresses of the order of hundreds of kg/cm^2, so that when measuring the static sensitivity of piezoceramic sound receivers, excess pressures of the order of atmospheric pressure can be used.

Any means can be used to create the excess pressure and both gaseous and liquid media can be employed. In our experiments the sound receiver was placed in a container which was filled with oil or water, and the excess pressure was created by means of air pressure at the free surface of the liquid. We employed an ordinary manually operated piston-type air pump. The excess pressure was measured by a precision manometer and the emf was measured with the aid of an Edelman-Luntz single string electrometer. This type of electrometer

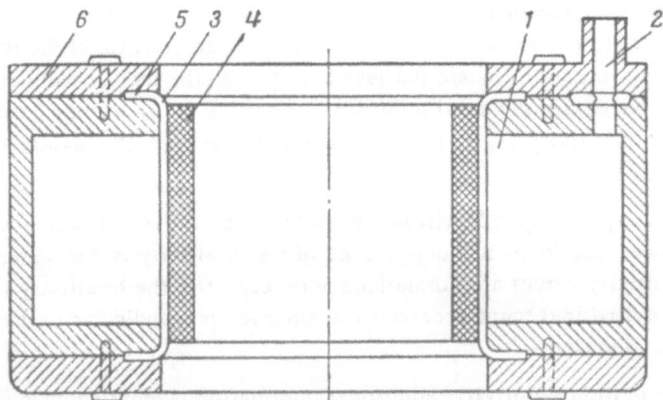

Fig. 9. Arrangement for creating a static pressure at the curved surface of a cylindrical piezoelectric element. 1) Cylindrical container; 2) connection to the pump; 3) rubber liner; 4) piezoelectric element; 5) edge of the liner; 6) flange.

possesses the advantage that its sensitivity can be adjusted by varying the potential difference between the stationary electrodes or by micrometric adjustment of the distance between them. In our experiments the values of the measured emf lay within the limits of a few tenths of a volt up to one volt.

The above-mentioned method was used not only to measure the static sensitivity of the assembled sound receiver but also served for preliminary checking of the sensitivity of the ceramic piezoelectric elements. However, in order to test the sensitivity of cylindrical piezoelectrical elements with different types of polarization, the apparatus shown in Fig. 9 was found to be more suitable. Excess air pressure is created in the cylinder 1 by means of an air pump connected to the nipple 2. The cylindrical piezoelectric element 4 is a close fit in the thin rubber liner 3. The edges of the rubber liner are turned back and clamped to the frame of the chamber by means of the flanges 6, thus ensuring an adequate hermetic seal. This method permits the application of excess pressure to the curved surface of a cylindrical specimen without having to mount the element in the assembled receiver.

The apparatus used to determine the static sensitivity of sound receivers can be and was used by us to test assembled piezoceramic sound receivers for mechanical strength and watertightness. In these cases we used excess pressures of the order of 10 atm.

3. Determination of the Frequency Dependence of the Electrical Impedance of Sound Receivers

The frequency characteristic of the electrical impedance of a sound receiver can be determined either as a combination of two frequency characteristics relating to the active and reactive components of the impedance, respectively, or as a combination of the two frequency characteristics relating to the modulus of the impedance and its phase angle, respectively. The latter method is more convenient since the frequency characteristic of the modulus of the impedance is an independent value, and in many cases it is sufficient to determine this characteristic alone.

The modulus of the electrical impedance was determined in the usual way by measuring the voltage at the sound receiver with the aid of an electron voltmeter at a given frequency and for a given current in the electrical circuit of the sound receiver. Thus, the frequency characteristic of the modulus of the impedance was determined either in the form of the frequency dependence of the voltage at the sound receiver for constant (independent of the frequency) current in the electrical circuit, or as the frequency dependence of the

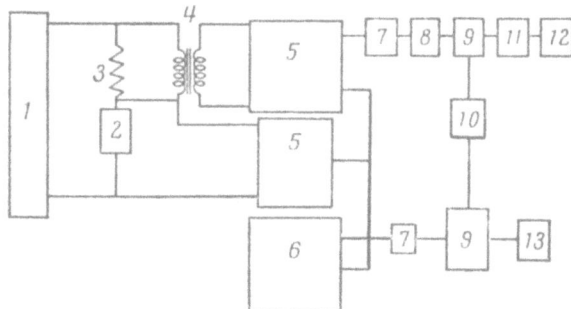

Fig. 10. Schematic diagram of the phase meter. 1) Audio-frequency voltage source; 2) sound receiver under test; 3) shunt resistance; 4) wide-band decoupling transformer; 5) balancing mixer; 6) first heterodyne; 7) first i.f. filter-amplifiers; 8) goniometric phase shifter; 9) mixer; 10) second heterodyne; 11) second i.f. filters; 12) amplifier limiters giving a differentiated output; 13) cathode-ray indicator.

current in the electrical circuit for constant (independent of the frequency) potential drop across the sound receiver. In order to avoid the difficulties normally associated with these measurements due to the presence of unconsidered measuring apparatus-ground capacities, thermoelectric milliammeters were used for current measurement.

The frequency characteristic of the phase angle of the electrical impedance was determined with the aid of a phasemeter represented schematically in Fig. 10.

The effect of the phasemeter circuit is to convert the voltages at the "potential" and "current" inputs of the phasemeter into a series of short pulses of identical and constant height generated at the instants when the corresponding voltages pass through zero. Both sets of pulses are fed to the vertical plates of a cathode-ray oscilloscope, the horizontal sweep of which is in synchronism with the pulse repetition frequency. The measurement process leads to the superpositioning on the oscilloscope screen of the two pulses (one of which corresponds to the series of pulses of the "current" channel, the other to the pulses of the "potential" channel). The superpositioning of the two pulses is achieved by rotating the rotor of the goniometric phase shifter, and the phase angle is read directly from the graduated scale showing the angle of rotation of the rotor which is graduated for both positive and negative angles.*

The range of operation of the phasemeter extends from 5 to 200 kcs and the phase angle is measured accurate to 1°.

The measurement of the electrical impedance of the sound receiver can also be used to estimate the electromechanical resonance frequencies of the receiver and thereby determine the working frequency range. Finally, if the resonance frequencies are very high, the upper limit of the working frequency range is determined by the decrease in sensitivity due to diffraction effects which arise when the dimensions of the receiver become comparable with the acoustic wavelength. In this case the sensitivity maxima corresponding to resonances lie outside the limits of the working frequency range and are not of great importance. However, in many cases for sound receivers used in air, we observe at higher frequencies, first, a certain decrease in sensitivity in comparison with the static sensitivity due to diffraction effects which is then followed by a new fairly sharp increase in sensitivity due to the first resonance. In this case the first resonance lies within the working frequency range. Finally, in the case of underwater sound receivers, for the most part the resonances occur simultaneously with or even before the reduction in sensitivity due to diffraction effects and the upper limit of the working frequency range is practically determined by the frequency of the first resonance. These considerations apply to wide-band sound receivers; in the case of narrow-band receivers the sensitivity maxima due to resonance effects are of prime importance.

Measurement of the electrical impedance of the sound receiver does not of course permit determination of the upper limit of the working frequency range if the latter is determined by diffraction effects, but at the same time does permit easy determination of the natural frequencies of the sound receiver. Pronounced resonances are easily determined by measurement of the modulus of electrical impedance as a function of frequency. Weakly expressed resonances are more easily determined by means of phase measurements.

Electrical measurements to determine the resonance frequencies were carried out not only for assembled sound receivers, but also for individual piezoelectric elements. These measurements provided the basis for judging the suitability of a given piezoelectric transducer for use in various types of sound receivers. Generally speaking, even in the case of underwater sound receivers, it is possible to determine the positions of the resonances by electrical measurement in air since the natural frequencies change only slightly as a result of immersion in water.

Generally speaking, when investigating sound receivers, it is preferable to carry out the electrical measurements before the acoustic measurements since the former are more easily carried out and together with the measurement of the static sensitivity give a rough idea of the sensitivity and working frequency. The acoustic measurements are best left until the later stages of investigation.

---

* The electron phasemeter described above was developed by Gotsak [183] at the Acoustic Institute of the Academy of Sciences of the USSR.

4. Determining the Frequency Characteristics of the Field Sensitivity of Acoustic Receivers

We employed a method based on the reciprocity theorem to investigate the frequency characteristics of acoustic receivers. The three-transducer method [184] is very suitable for measuring the sensitivity of spherical and cylindrical piezoceramic sound receivers. We also carried out measurements in water in a nonanechoic tank employing pulsed radiation. In this case it is convenient to locate all three transducers, i.e., the investigated sound receiver, the reciprocal transducer (transmitter-receiver), and the auxiliary transducer-radiator in the same horizontal plane separated by equal distances, i.e., at the angles of an equilateral triangle. If the reciprocal transducer and the transducer-radiator are nondirectional in the horizontal plane (spherical or cylindrical transducers), there is no necessity for mutual displacement or rotation of the transducers during the process of measurement. The arrangement used to mount the three transducers is shown schematically in Fig. 11. The transducers 1 are secured to light duralumin holders 2. In order to prevent the transmission of energy from the transmitting transducer to the receiver transducers in the form of vibrations transmitted along the mounting system, the holders are fitted with rubber shock absorbers 3. The tubular holders are made in sections so that by raising vertically the lower telescopic member of the holder it is possible to vary the depth of immersion of the transducers. All three vertical tubular members are secured to horizontal rods 4 by means of the sliding sleeves 5. This arrangement permitted accurate locating of the transducers at the corners of an equilateral triangle with the required length of side. The horizontal rods were secured to a hollow cylindrical member 6 which also acted as a float to support the apparatus when immersed in the tank.

The pulse measuring system is shown schematically in Fig. 12. The pulse modulator 2 generates input pulses with a pulse length τ_1 and a high frequency duty cycle related to an instant of time taken as zero. The modulator is controlled by the timer 3 by means of rectangular pulses of pulse length τ_1 related to the same time reference.

The oscilloscope 5 is used to measure the pulse current and the oscilloscope 6 to measure the pulse voltage. The receiver transducer 8 is located at a distance L from the transmitter transducer 7. The pulse selector 10 opens the receiving channel for a selected time interval τ_2 with a certain time delay with respect to zero time. The selector is controlled by rectangular pulses of duration τ_2 generated by the timer 3 with a variable time delay which in our experiments was chosen equal to L/c, i.e., the time taken by the pulse to travel from

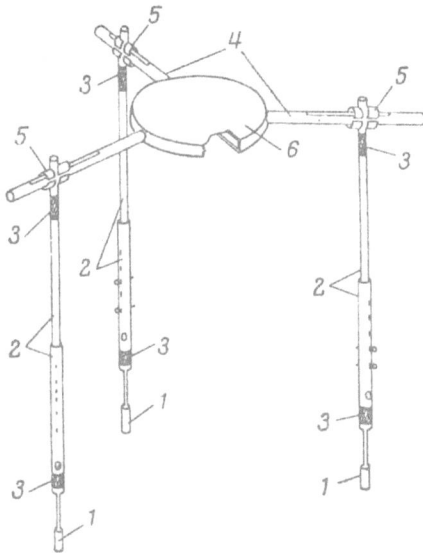

Fig. 11. Arrangement for mounting the three transducers
for calibration by the reciprocity method.

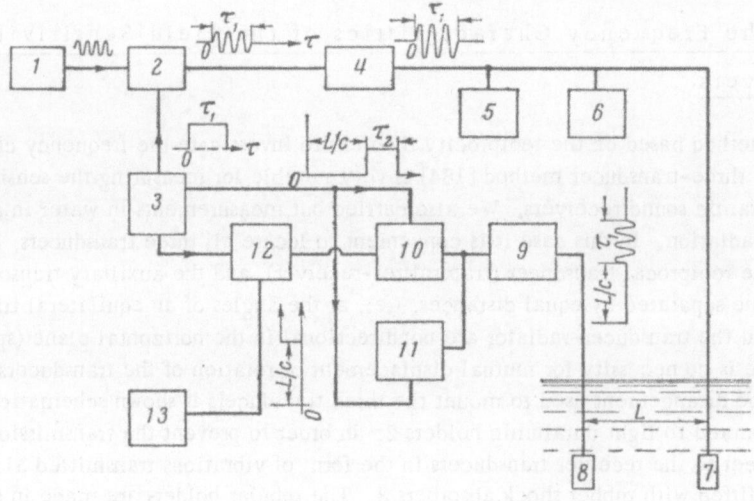

Fig. 12. Schematic diagram of the pulse system for transducer calibration
by the reciprocity method. 1) Master oscillator; 2) pulse modulator; 3)
timer; 4) final amplifier; 5. 6) oscilloscopes; 7) transmitting transducer;
8) receiving transducer; 9) wide-band amplifier; 10) pulse amplifier; 11)
dual-beam oscilloscope; 12 memory unit; 13) measuring instrument.

the transmitter to the receiver. In most cases the receiving channel was open for a period selected as 0.1 msec; in this case by varying the time delay, it was possible to observe the form of the received pulse. In order to obtain an overall picture of the pulse form in reverberation processes in the tank we used a dual-beam oscilloscope 11, the upper trace giving the amplitude of the signal with respect to time and the lower trace the position of the gate pulse indicating the time and duration of opening of the receiving channel. The gate pulse at the oscilloscope 11 is obtained by feeding a corresponding voltage from the timer 3. The high-frequency pulse voltage from the selector 10 is amplified and fed to the memory unit 12 which delivers a maintained constant voltage proportional to the pulse amplitude. This constant voltage is measured by means of the pointer-type measuring instrument 13. The timer carries out a cycle of operations with a variable duty cycle. The pulse repetition frequencies used in our experiments varied from 1 to 50 cycles and the pulse duration from 2 to 20 msec. During cyclic operation the memory unit 12 was cleared immediately prior to commencement of the next cycle. The instant of clearing was controlled by a pulse transmitted from the timer.

The time required to clear the memory unit and generate a new voltage following the arrival of the next pulse was so short as to have practically no effect on the accuracy of the indicating instrument 13.

The oscilloscopes 5 and 6 were calibrated with the aid of a substitution voltage obtained from the auxiliary generator at the corresponding frequency and measured by an electron voltmeter. Exactly the same method of direct electric calibration of the receiver channel was carried out under continuous operation with the aid of a calibrating voltage from the auxiliary generator with attenuator.

The switching of the receiving and transmitting channels of the measuring system to the corresponding transducers was achieved by means of a commutation arrangement which made it possible to carry out the following three series of measurements required to determine the sensitivity by the three-transducer method.

1. A sound field is generated by the auxiliary transmitter. Measurements are made of the current through the transmitter and the no-load voltage V_x developed by the investigated sound receiver.

2. The receiving channel is switched to the reciprocal transducer and the no-load voltage V developed by the reciprocal transducer operating as a receiver is measured for the same current through the auxiliary transmitter.

Fig. 13. Apparatus for locating and rotating transducers during tank measurements.

3. The transmitting channel is now switched to the reciprocal transducer which now generates a sound field. The current I through this transducer is measured and the no-load voltage V'_x developed in this case by the investigated sound receiver is measured with the aid of the receiving channel.

The sensitivity of the investigated sound receiver can now be determined from the above-mentioned measurements by means of the formula

$$E_x = \sqrt{H \frac{V_x}{V} \frac{V'_x}{T}},$$

where H is the reciprocity constant which in the given case we assume equal to

$$|H| = \frac{2R\lambda}{\rho c}, \tag{1}$$

where R is the distance between the transducers, c is the velocity of sound in water, λ is the acoustic wavelength corresponding to the frequency at which the measurements are made, and ρ is the density of water.

The three-transducer method is not very satisfactory for measuring the sensitivity of sound receivers whose directivity characteristics are axially symmetric, e.g., receivers with plane diaphragms, since this method involves variation in the angle of orientation of at least the investigated transducer during the period of measurement. Therefore, in order to determine the frequency characteristics of the sensitivity of sound receivers with marked axial sensitivity, we employed a pulse method of measurement with reflection of the signal from the free surface of the water. Naturally such a method can only be used for dual-purpose transducers, i.e., for sound receivers which can also be used as transmitters.

The tank in which the experiments were carried out was equipped with a mobile system for locating the transducer at any desired point (seen in the background of Fig. 13) and an automatically operated rotating device for determining the directivity characteristics of transducers (right-hand side of Fig. 13). By means of this system the transducer could also be located at any depth in the tank with its emitting-receiving surface strictly parallel to the free surface of the water.

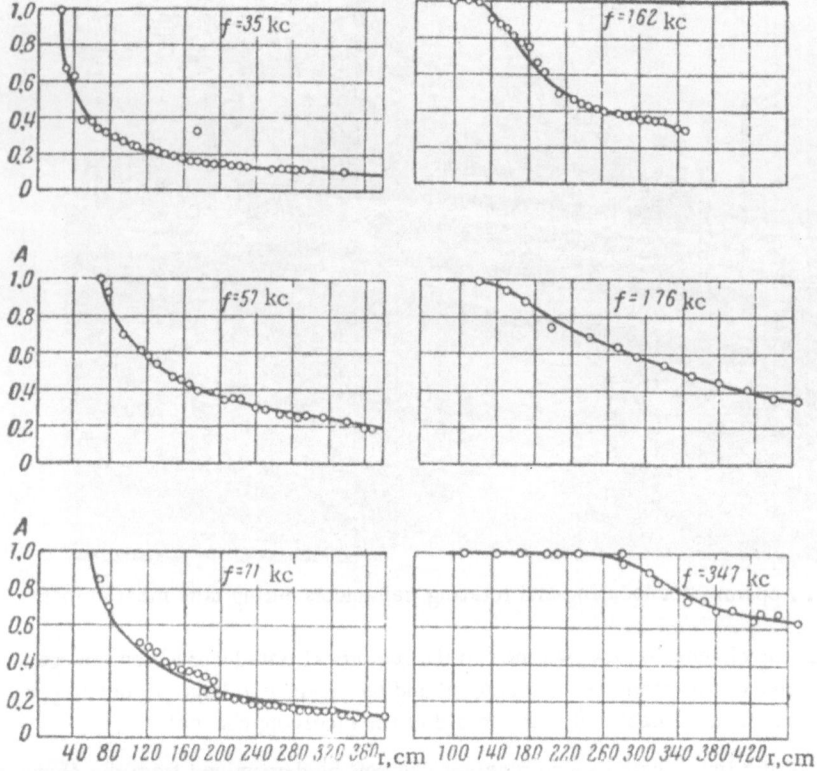

Fig. 14. Dependence of the amplitude of the received signal on depth of immersion during calibration of plane transducers with the aid of reflection from the free water surface.

The measurements were carried out as follows: 1) the current through the transducer I was measured with the transducer operating as a transmitter; 2) the pulse reflected from the water surface was picked up by the sound receiver and the emf E developed by the transducer acting as a receiver was measured. On the basis of these measurements the axial sensitivity of the sound receiver (under no-load conditions) was determined in accordance with the formula

$$E_x = \sqrt{H \frac{E}{I}},$$

where H is the reciprocity constant which in the given case is equal to

$$H = \frac{4r\lambda}{\rho c},\tag{2}$$

where r is the distance from the sound receiver to the free water surface, λ is the wavelength, c is the sound velocity, and ρ is the density of water.

It is possible simultaneously to determine the sensitivity of the transducer operating as a transmitter with the aid of the formula

$$R_x = \sqrt{\frac{1}{H} \frac{E}{I}}$$

(which employs the same notation as above).

The reciprocity constants given by (1) and (2) only apply when the distances R and r considerably exceed the axial extent of the projector zone of the transducer at the given frequency. In order to investigate the sound field created by the transducer acting as a transmitter we investigated the dependence of the amplitude of the reflected signal on the depth of immersion r of the transducer (Fig. 14). The transducer consisted of a circular plate 10 cm in diameter enclosed in a housing which was free from radiation on the reverse side.

It is clearly shown that so long as the distance from the transmitting surface of the transducer to the free water surface is less than twice the extent of the project or zone of the transducer at the given frequency, the strength of the reflected signal at the receiver remains constant; and only when this distance is increased, does the signal strength at the receiver begin to decrease and gradually approximate to the spherical law. At relatively low frequencies (up to 100 kcs) the signal amplitude begins to decrease in accordance with the spherical law at relatively small transducer immersion depths (of the order of tens of cm), and consequently when measuring the sensitivity in this frequency region it is possible to use the reciprocity constant given by (2). However, because of the restricted depth of the tank, at higher frequencies it is not possible to achieve sufficiently large transducer immersion depths, and we are constrained to carry out the measurements within the limits of the projector zone of the transducer. In this case the transmitted (and correspondingly, received) wave can be assumed plane and the reciprocity constant is then given by

$$H = \frac{S}{\rho c},$$ (3)

where S is the area of the transmitting surface of the transducer (or in other words, the cross-sectional area of the sound beam in the projector zone), c is the velocity of sound, and ρ is the density of water.

The use of (3) involves the assumption that the transducer is acoustically perfectly rigid. If this condition is not fulfilled in practice, a correction which takes into account the acoustic impedance of the transducer must be applied. However, this question will be examined in greater detail when methods of investigating sound receivers in tubes and solid delay lines are described.

When the sensitivity is measured by the three-transducer method or by the method of reflection from the water surface at distances exceeding the extent of the projector zone, we obtain the sensitivity of the receiver in a free field. When the measurements involving reflection are carried out within the limits of the projector zone, we actually obtain the pressure sensitivity and introduce a diffraction correction equal to 2 which corresponds to a perfectly rigid transducer whose diameter is a number of times greater than the wavelength in the medium.

5. Determination of the Frequency Characteristics of the Pressure Sensitivity of Sound Receivers

In the case of transducers with plane receiving-transmitting surfaces it is convenient to use methods of sensitivity determination based on the use of waveguides of restricted cross section. Metal tubes filled with a liquid (e.g., water) are most frequently used for this purpose [185]. In order to test sound receivers used in air, gas-filled (air or hydrogen) tubes must be used [186]. We used a vertical tube filled with water; we also used the same measuring apparatus and the reflection method of sensitivity determination employed in the case of the tank measurements. Naturally, measurements in a tube filled with water involve a number of possible sources of error. Firstly, we must take into account the compliance of the tube walls and select a wall thickness such that the dispersion frequency region due to compliance lies outside the limits of the frequency range of the measurements. At frequencies up to 30 kcs we used a brass tube with an internal diameter of 30 mm and a wall thickness of 15 mm.

When carrying out measurements in a tube it is only possible to obtain the pressure sensitivity of the sound receiver; and if it is necessary to determine diffraction corrections, additional measurements must be carried out in a free field. However, in many cases measurement sound receivers are in fact used in tubes and therefore pressure sensitivity possesses an interest in its own right.

A second possible source of error associated with measurements in water-filled tubes arises from the presence of air bubbles trapped at the tube walls when the tube was filled with water. Thus, degassed water should be used and the tube should be filled as slowly as possible.

The disadvantages of liquid-filled waveguides can to a considerable extent be avoided by the use of a solid waveguide constructed from a material which is a poor sound absorber, e.g., aluminum, bronze, or other similar metals. So far as we know, we were the first to use solid waveguides for the calibration of electro-acoustic transducers.

We selected a waveguide in the form of a duralumin rod of circular cross section which was fairly small and easily rotated. In addition duralumin is easily machineable and it was possible to obtain good mechanical contact between the plane surface of the transducer diaphragm and the end surface of the waveguide. These surfaces were ground and lapped in oil before the commencement of the experiment.

The use of solid waveguides makes it possible to investigate sound receivers of even very low sensitivity since, in this case, the acoustically rigid transducer provides a much better match with the medium than in the case of a water-filled waveguide.

Measurements in both liquid and solid waveguides were carried out by the pulse method using the signal reflected from the free end of the waveguide. We employed the same apparatus used in the measurements carried out in the tank utilizing reflection from the free water surface.

As previously stated, the waveguide method of transducer calibration can only be used to determine the pressure sensitivity of the sound receiver. By pressure sensitivity we understand the ratio of the emf developed by the transducer acting as a receiver to the pressure actually existing at the transducer diaphragm. Naturally, the pressure sensitivity is independent of the parameters of the medium used during the measurements. Thus, in order to measure the pressure sensitivity we can use various types of waveguide, including solid waveguides in which case only longitudinal waves are used. At low frequencies the pressure sensitivity of a transducer co-incides with the static sensitivity.

We shall now derive the relationships required for the calculation of the pressure sensitivity on the basis of experimental data obtained with the aid of liquid or solid waveguides.

The pressure sensitivity of the transducer will be expressed in the form $M = E/p_{\Phi}$, where p_{Φ} is the actual pressure exerted at the receiver diaphragm. In addition we introduce the secondary concept of "field sensitivity" for simultaneous operation of the receiver and waveguide [187]. This sensitivity is determined as the ratio of the emf developed by the sound receiver to the sound pressure which would exist at a given cross section of the waveguide if the latter were of infinite extent. The field sensitivity is expressed as $M' = E/p_0$, where p_0 is the traveling-wave sound pressure. The pressure sensitivity M and the field sensivity M' are related by the expression

$$M = \frac{M'}{1 + \beta} ,$$

where β is the coefficient of reflection at the receiver surface.

Transducer sensitivity measurements carried out by the waveguide method are essentially similar to those carried out by the method of reflection from the free water surface. The effective value of the pulse current I_1 is measured at the instant of pulse transmission, and the emf E_1 developed by the sound receiver is measured at the instant arrival of the reflected pulse at the receiver. The sensitivity M'_1 is calculated from the formula

$$M'_1 = \sqrt{\frac{E_1}{I_1} H} ,$$

where H is the reciprocity constant equal to $S/(\rho_1 c_1)$ (S is the transverse cross section of the waveguide equal to the area of the diaphragm; $\rho_1 c_1$ is the wave impedance of the waveguide medium). Thus, the pressure sensitivity is equal to

$$M = M_1 = \frac{1}{1+\beta} \sqrt{\frac{E_1}{I_1}} H.$$

In the case of plane waves

$$\beta = \frac{z-1}{z+1},$$

where Z is the input mechanical impedance of the sound receiver.

Thus, we finally obtain

$$M = M_1 = \sqrt{\frac{E_1}{I_1} \frac{S}{\rho_1 c_1} \frac{(z+1)^2}{(z-1)^2}}.$$

Our formulae contain the input mechanical impedance of the investigated transducer. Therefore, in order to determine the sensitivity we must also determine experimentally the frequency dependence of the input impedance. The latter is a complex quantity so that it is necessary to determine the frequency characteristics of its active and reactive components. This can be done by measuring the modulus of the coefficient of reflection and the phase displacement between the reflected and incident waves (using the investigated transducer as a reflecting object). For this purpose it is located at the free end of the same liquid or solid waveguide used to determine the sensitivity. An additional transducer is placed at the other end of the waveguide to transmit and receive the signal pulse.

Generally speaking, we measured the mechanical impedance of the transducer using a duralumin waveguide, i.e., we measured the quantity $Z/\rho_1 c_1$ where $\rho_1 c_1$ is the characteristic impedance of duralumin. Conversion to the value of the impedance in water is easily achieved by means of the formula

$$\frac{z}{\rho c} = \frac{z}{\rho_1 c_1} \frac{\rho_1 c_1}{\rho c},$$

where ρc is the characteristic impedance of water.

The method of measuring the mechanical impedance has been described in detail by Ageeva [188]. In our measurements we used the electronic channel of the Ageeva apparatus which was kindly placed at our disposal.

If we experimentally determine the modulus of the coefficient of reflection $|\beta|$ and the plane displacement φ between the incident and reflected waves, then the active and reactive components of the input mechanical impedance and its modulus can be obtained from the known formulae

$$x = \frac{1-\beta}{1+\beta^2 - 2\beta \cos \varphi},$$
$$R = \frac{2\beta \sin \varphi}{1+\beta^2 - 2\beta \cos \varphi},$$
$$|z| = \sqrt{R^2 + x^2}.$$

The data obtained with the use of a solid waveguide (in our case a duralumin rod) were compared with corresponding data obtained with a water-filled tube, satisfactory agreement being obtained.

The method of sensitivity measurement using solid waveguides appears to be particularly suitable for investigation of transducers designed for use in connection with sound transmission in solids, e.g., sound transmitters and receivers used in flaw detectors.

6. Determining the Directivity Characteristics of Sound Receivers

Directivity characteristics were chiefly determined in open water and the aid of the automatic system used on the floating laboratory of the Volzhskii scientific research station of the Acoustic Institute of the Academy of Sciences of the USSR.

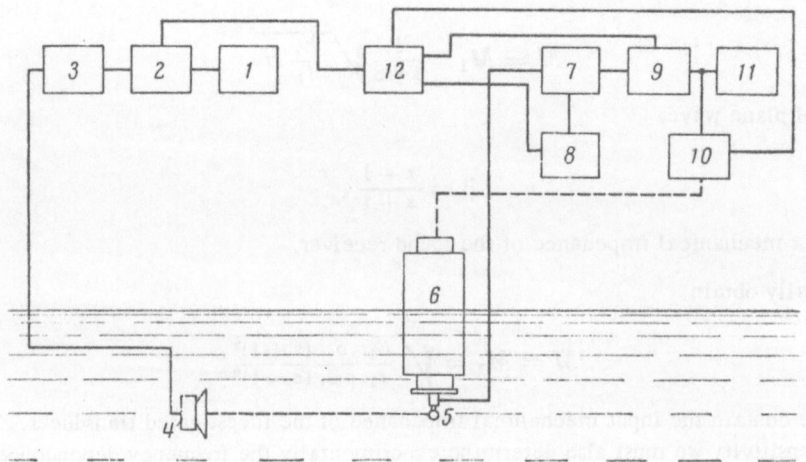

Fig. 15. Schematic diagram of the arrangement used for automated recording of the directivity characteristics of sound receivers. 1) Master oscillator; 2) pulse modulator; 3) final amplifier; 4) underwater sound transmitter; 5) investigated sound receiver; 6) rotating system; 7) wide-band amplifier; 8) dual-beam oscilloscope; 9) memory unit; 10) directivity recorder; 11) output electronic voltmeter; 12) timer.

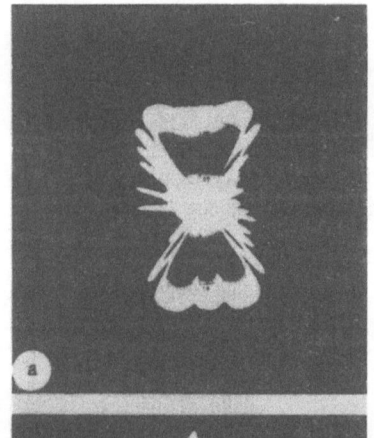

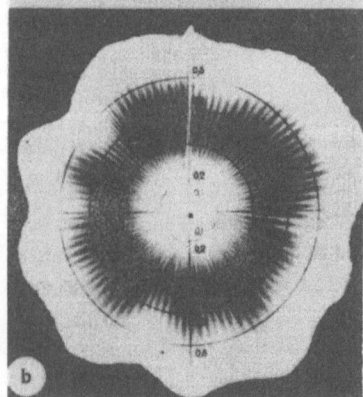

Fig. 16. Transducer directivity characteristics.

This system consists of electrically operated lifting and rotating systems, a transmitting channel, and a receiving-recording channel.

The arrangement is shown schematically in Fig. 15.

The directivity recorder 10 consists of a cathode-ray tube with a large-diameter screen on which the directivity characteristics of the investigated sound receiver are displayed and recorded by photography. Rotation of the sweep system is achieved by means of electric coupling between the recorder and the elevating-rotating system 6.

The elevating-rotating system 6 not only automatically rotates the sound receiver 5, but also corrects for depth. The working frequency range of the electronic channels of the measuring system extends from 10 to 500 kcs. The pulse duration can be varied from 0.2 to 2.0 msec and the pulse frequency from 1 to 50 cycles. The timer 12 opens the receiving channel for a period equal to or less than the pulse duration, depending on the time delay, thus eliminating the effect of unwanted reflections. The dual-beam oscilloscope 8 permits observation of the effective signals and reverberations in the water space and also, with the aid of a marker gate, selection of the required instant of gating the receiving channel.

The delay time can be varied within limits of 0.2 to 20 msec which enables any separation of the transmitter and investigated sound receiver within the limits of the length of the floating laboratory. The speed of rotation was set at approximately one revolution per minute which not only allowed visual observation of the directivity characteristics on the recorder 10, but also recorded with adequate detail the directivity characteristics of highly directional transducers possessing characteristics with aperture angles of a few degrees.

In our experiments the distance between the transmitter 4 and sound receiver 5 was varied between the limits of 0.5 and 7 m. The transmitter was secured by a rod to a carriage which traveled along rails laid in a channel in the deck of the floating laboratory. Figure 16 shows photographs of directivity characteristics obtained with the arrangement described above.

7. Sequence of Operations Used for the Investigation of Sound Receivers

On the basis of our experience we can recommend the following sequence of measurements.

First measure the capacity and static sensitivity of the piezoelectric element to be used in the sound receiver; also measure the loss angle and insulation resistance. After assembling the sound receiver, approximately determine the working frequency range of the sound receiver; this requires determination of the corresponding frequency dependence of the electrical impedance (in air). Now check the static sensitivity, capacity, loss angle, insulation resistance, etc., of the assembled sound receiver. The results obtained are compared with the corresponding data for the isolated piezoelectric element in order to establish the quality of the assembly.

If the assembly has been carried out correctly, the frequency characteristics of the sound receiver are now determined and the working frequency range established more accurately. Depending on the purpose for which the sound receiver was designed, measurements are now carried out by one of the methods described above. The frequency characteristic is naturally determined with respect to a given reference direction (e.g., frequency characteristic of the axial sensitivity).

Finally, for a given set of frequencies the directivity characteristics are determined and normalized with respect to the sensitivity in the reference direction.

The given sequence of operations generally guarantees that the required data are obtained with a minimum of measurement effort.

CHAPTER III

NONDIRECTIONAL WIDE-BAND SOUND RECEIVERS

1. Establishment of the Problem. First Mock-Up of the Nondirectional Sound Receiver

The problem of the design of miniature nondirectional sound receivers was submitted to the acoustic laboratory of the P. N. Lebedev Institute of Physics, Academy of Sciences of the USSR, in connection with investigation of the spatial structure of sound fields. Such investigations had previously been carried out with the aid of sound receivers containing piezoelectric crystal elements of cubic form. However, in 1950 results obtained by a group of investigators working under the direction of V. S. Grigor'ev showed that the directivity characteristics of such sound receivers differed considerably from spherical even when the dimensions of the piezoelectric element were much smaller than the acoustic wavelength in the ambient aqueous medium. These miniature sound receivers indicated marked directional properties even when their dimensions were 10 to 15 times smaller than the acoustic wavelength.

For example, Fig. 17 shows the directivity characteristic of a "point" sound receiver with an ethylenediamine tartrate element with dimensions of $2 \times 2 \times 2$ mm^3 at a frequency of 50 kc, i.e., for a wavelength in water of 3 cm. The element was enclosed in a thin-walled cylindrical Plexiglas container with a diameter of 40 mm. The cylinder was filled with transformer oil in order to insulate the element from the surrounding aqueous medium. The directivity characteristic displays troughs which extend right down to zero. The same behavior was also observed at other frequencies.

Similar experiments were also carried out with miniature piezoelectric elements of polarized barium titanate ceramic. Figure 18 shows the directivity characteristic of a piezoelectric ceramic sound receiver containing a cubic element with dimensions of $1 \times 1 \times 1$ cm^3. In order to eliminate any possible effect on the directivity characteristic due to the insulating container, the piezoelectric element was coated with water-resistant lacquer and the directivity characteristic was determined in the absence of a container. The photographs show that dipole and quadrupole characteristics are more clearly in evidence than all-round reception. This is presumably due to the fact that the hydrostatic coefficient for barium titanate ceramic, $d_{\text{hyd. coeff}} = d_{33} + 2d_{31}$, is much smaller than the piezoelectric coefficient in the direction of polarization d_{33} (d_{33} and d_{31} possess opposite signs).

Almost all known piezoelectric crystals (with the exceptions of lithium sulfate, tourmaline, L-rhamnose) possess crystallographic axes in whose direction, or at a given angle to this direction, the piezoelectric properties of the crystal are expressed more strongly than the piezoelectric properties determined by the hydrostatic coefficient. Therefore, it is apparently not easy to design a sound receiver with a crystal element in which the emf developed by the crystal is proportional to the hydrostatic coefficient.

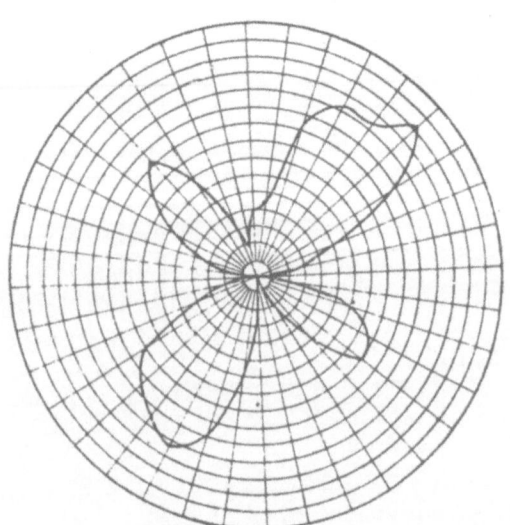

Fig. 17. Directivity characteristic of a "point" ($2 \times 2 \times 2$ mm^3) ethylenediamine tartrate sound receiver at a frequency of 50 kc.

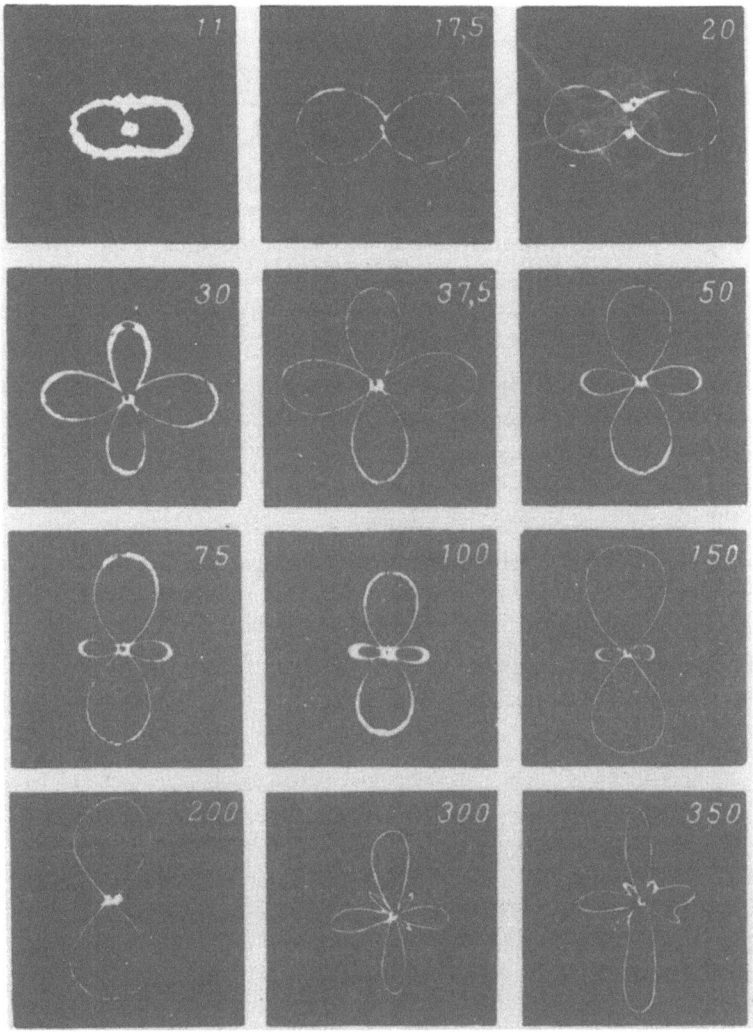

Fig. 18. Directivity characteristic of a sound receiver consisting of a cube of barium titanate ceramic with dimensions of $10 \times 10 \times 10 \text{ mm}^3$. The figures in the top right-hand corner of the diagrams indicate the frequency in kc.

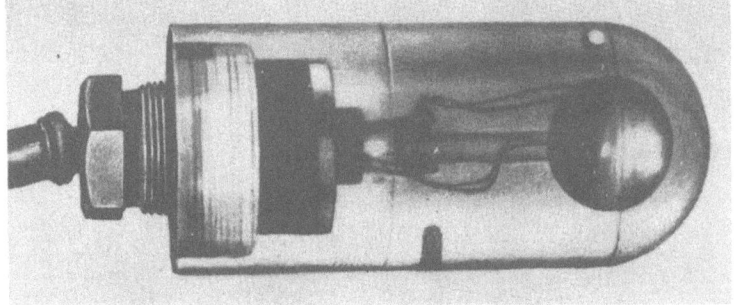

Fig. 19. First model of a nondirectional, spherical sound receiver containing a barium titanate ceramic element.

TABLE 8

Frequency, kc	0 (static measure-ments)	25	30	45	60	75	100
Sensitivity, μV/bar	1.7	0.6	0.88	1.6	1.7	1.25	1.45

Figure 18 shows that the directional properties steadily decrease with decrease in frequency. Therefore, in principal, piezoelectric elements with even smaller dimensions are required. However, this would inevitably mean a decrease in sensitivity and capacity of the piezoelectric element which would be most undesirable. A better approach is to create piezoelectric elements whose forms possess radial symmetry and this, in fact, has become possible with the introduction of piezoceramics.

In 1950, following the suggestions of N. N. Andreev and V. S. Grigor'ev, the Acoustic Institute of the Academy of Sciences of the USSR prepared spherical, radially-polarized elements of barium titanate ceramic which enabled the production of the first prototype nondirectional sound receiver for use at ultrasonic frequencies (up to 100 kc) [44]. Tests of this first spherical-type sound receiver demonstrated the importance of research in this new direction.

Figure 19 shows the first model of a nondirectional spherical sound receiver of barium titanate ceramic. The receiving element consisted of two hollow hemispheres of barium titanate ceramic glued together. Fused silver electrodes were deposited on the internal and external surfaces of the hemispheres. The lead from the internal electrode was brought out through a small ceramic tube which also served as the mounting for the sphere.

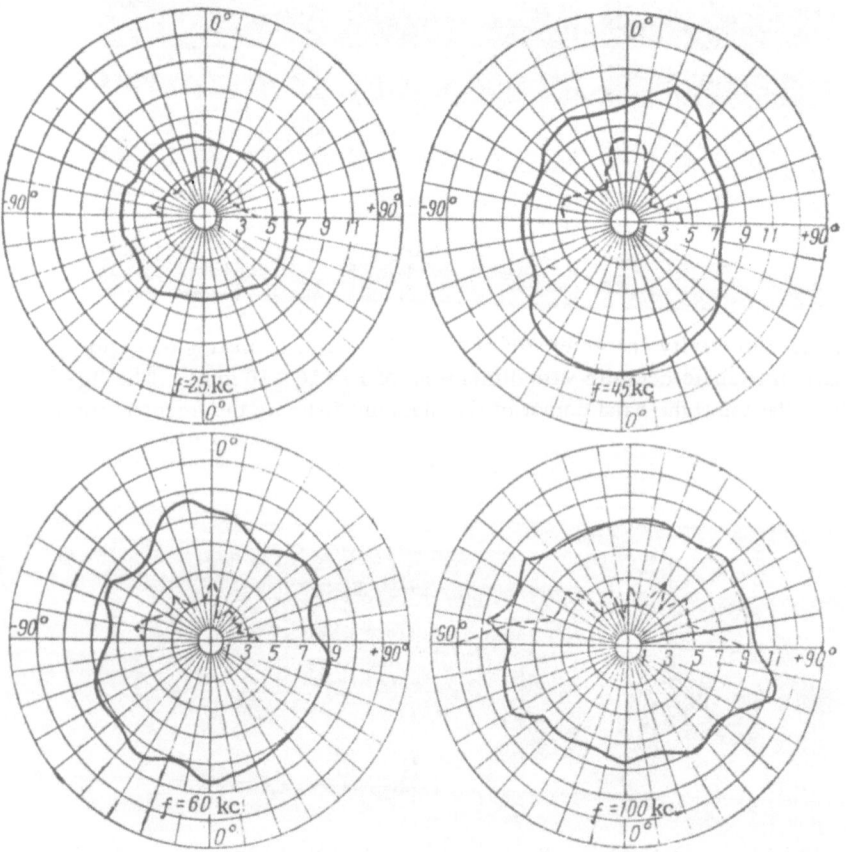

Fig. 20. Directivity characteristics of model of the first prototype spherical sound receiver.

The spherical, radially-polarized element was placed in a Perspex container filled with transformer oil. The sensitivity of early models of this type of sound receiver with external diameters of the sensitive element of 14 and 16 mm and a wall thickness of 2 mm was fairly small. The static sensitivity was approximately 1.7 µV/bar; the field sensitivity varied from 0.6 to 1.7 µV/bar.

The capacity of piezoelectric elements with external diameters of 14 and 16 mm was 1000 and 1400 $\mu\mu f$. The frequency dependence of the sensitivity can be estimated from the data given in Table 8 for one of these sound receivers.

Examples of the directivity characteristics of the first prototype model of the spherical sound receiver are shown in Fig. 20. The continuous line gives the directivity characteristic of the receiver in a plane in which the full symmetry of construction is maintained; the broken line gives the directivity characteristic in the plane passing through the mounting of the spherical element. In what follows we shall arbitrarily designate these planes as horizontal and vertical, respectively.

Even in the first models the directivity characteristic in the horizontal plane could be regarded as satisfactory in comparison with corresponding characteristics for sound receivers using piezoelectric crystal elements. In the horizontal plane, in which there is no constructional asymmetry. the nonuniformity of the directivity characteristic at high frequencies amounted to approximately 5 db. In the vertical plane, the directivity characteristics of the first experimental models could not be regarded as satisfactory. The deviation from a circular directivity characteristic in this plane is due to the effects of the sphere mounting and the container.

Despite the fact that the first nondirectional sound receivers described above did not give completely satisfactory results, the data obtained was sufficient to show that research was undoubtedly proceeding in the right direction. We subsequently improved the performance of spherical sound receivers and also designed sound receivers which were nondirectional in one plane only (horizontal plane), i.e., sound receivers possessing not spherical but specific cylindrical symmetry of the sensitive element. In many regions of application such cylindrical sound receivers are equally as important as spherical receivers.

In subsequent years, spherical sound receivers with barium titanate ceramic elements found wide application as measuring hydrophones [189]. These hydrophones, produced in the Soviet Union, embraced a frequency range from 50 to 100 kc. Similar measurement sound receivers were produced by the Rostov Institute for Technology and Machine Construction to the design of the Rostov State University [190]. Cylindrical sound receivers with barium titanate ceramic elements were also successfully used in the search for oil deposits [191].

As stated above our original efforts were directed at problems associated with the investigation of the structure of sound fields. At first we concentrated on the development of relatively small (miniature) sound receivers with a high upper frequency limit and comparatively low sensitivity, and subsequently in an attempt to increase the sensitivity of sound receivers designed for various types of investigation at lower frequencies, we gradually increased the dimensions up to diameters of 5 cm for spherical receivers and 12 cm for cylindrical receivers. In recent years, a need has arisen for subminiature sound receivers. We have developed spherical and cylindrical sound receivers with sensitive elements only 2 mm and 2-3 mm in diameter, respectively. These receivers are used to investigate the structure of high frequency acoustic fields in liquid media. As a special item we developed a cylindrical element for a sound receiver for insertion into the cardiovascular system for medical diagnosis.

Romanenko [192] developed spherical-type miniature sound receivers with diameters of 0.2-0.3 mm for investigating pressure distribution at the front of shock waves [193].

The foreign literature contains very few papers concerning the use of piezoelectric ceramics in measurement sound receivers. We know only the paper by Ackerman and Holak [48] which describes subminiature sound receivers 0.5 mm in diameter with barium titanate ceramic elements operating in the frequency range from 10 kc to 1 Mc. The great majority of foreign papers concerning ceramic transducers deal with their properties as transmitters. Publications of this type will be found in the review paper by Bradfield [53] and in the paper by Johnston and Wertz [52], etc.

2. Calculating the Static Sensitivity of a Sound Receiver with a Thin-Walled Spherical or Cylindrical Shell

The sensitivity of sound receivers of the simplest construction with rod- or package-type transducer elements is determined in the absence of loading by the g-coefficient which is proportional to the ratio of the d-coefficient and the dielectric constant ε. In the case of a polarized ceramic we have three possible variants shown in Fig. 21. In case (a) the direction of the external force (and consequently the direction of longitudinal deformation) coincides with the direction of the polarization axis z; in cases (b) and (c) in direction of the external force F is perpendicular to the polarization axis z. When calculating the sensitivity of the transducer we must use the d_{33} coefficient in case (a) and in cases (b) and (c), d_{31} and d_{32}, respectively. Since in the case of a polarized piezoceramic with axial symmetry $d_{31} = d_{32}$, the variants (b) and (c) are identical. If the dimension of the piezoelectric element in the direction of polarization is indicated by l, the no-load sensitivity for case (a) is given by the formula

$$\frac{E}{p_0} = g_{33} l ,$$

and for cases (b) and (c) by

$$\frac{E}{p_0} = g_{31} l = g_{32} l,$$

where p_0 is the pressure at the surfaces perpendicular to the direction of the external force F.

Tables 4 and 5 show that the values of the parameters g_{33} and g_{31} for ceramic piezoelectric materials are inferior to the corresponding values for other piezoelectric materials. For example, barium titanate ceramic is 5 times less sensitive than quartz, 10 times less sensitive than potassium phosphate, and 39 times less sensitive than Rochelle salt. Therefore, piezoceramic elements in the form of assemblies in which the pressure is incident on one of the lateral surfaces do not give satisfactory results. Therefore, the guiding principle in the design of sensitive piezoceramic sound receivers is to achieve amplification of the voltage set up in the transducer, by some type of mechanical transformation, and corresponding development of the external surface of the sound receiver on which the sound pressure is incident. In particular, this can be achieved by using a thin-walled shell constructed from a piezoceramic material. Naturally, we must also try to increase the effective dimension l, i.e., the distance between the electrodes in rod- or assembly-type piezoelectric elements; however, this introduces either a decrease in the capacity of the element or an increase in size which in most cases is undesirable.

The sensitivity of a sound receiver with mechanical transformation can, generally speaking, be expressed by the formula

$$\frac{E}{p_0} = K g l,$$

where K is the coefficient of mechanical transformation. The effective utilization of the piezoelectric material increases with increase in K.

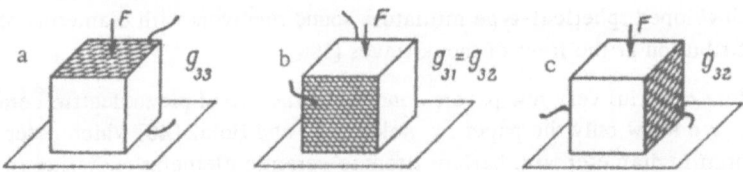

Fig. 21. Three simplest variants of a plane polarized ceramic piezoelectric element.

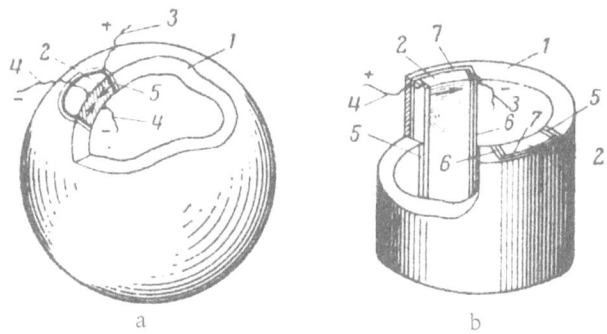

Fig. 22. Sound receivers with (a) spherical, (b) cylindrical receiver surfaces. 1) Thin-wall receiver shell; 2) ceramic; 3, 4) electrodes and leads; 5, 6, 7) insulating layer.

In designing sound receivers with spherical or circular directivity characteristics it is natural to use a spherical or cylindrical shell for mechanical transformation. In this case it is not absolutely necessary that the whole of the shell shall consist of piezoceramic material. The piezoelectric element can be embedded in the spherical or cylindrical shell as, for example, shown in Fig. 22. In this type of construction, the functions of mechanical transformation and electromechanical conversion are divided among special elements of the sound receiver.

We shall determine the coefficient of mechanical transformation for spherical and cylindrical shells. For the sake of simplicity we assume that the wall thickness is so small in comparison with the external diameter that the normal stresses in the shell acting in a radial direction can be neglected and assume that the external pressure acting at the surface of the shell creates normal stresses only acting in a tangential direction in each diametral cross section of the shell. The ratio of the tangential mechanical stress σ_T to the sound pressure acting at the surface of the shell, i.e., σ_T/p_0, determines the coefficient of mechanical transformation in any given case. We shall assume that the sound pressure is identical over all the elements of the external surface of the shell, which means that the shell diameter is small in comparison with the wavelength. Under these conditions it is possible to neglect the tangential stresses which otherwise would be related with shear deformations of the shell.

Let us consider first the case of a thin-walled spherical shell of radius R, the external surface of which is subjected to a given excess pressure p_0. Let the sphere be dissected by two mutually perpendicular great circles (Fig. 23). Then at the point of intersection of the diametral cross sections of the sphere (i.e., at any point on the sphere) we can determine the mechanical tangential stresses in the shell σ_T and σ_ψ which are

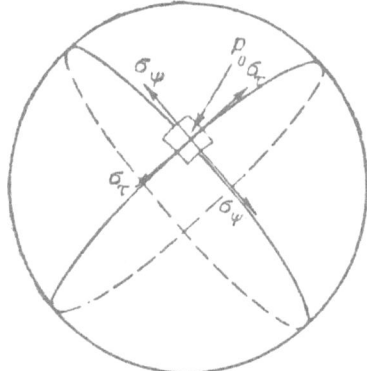

Fig. 23. The case of a thin-walled spherical shell.

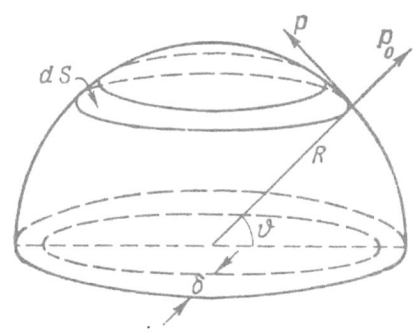

Fig. 24. Calculation of the total force F acting on the hemisphere.

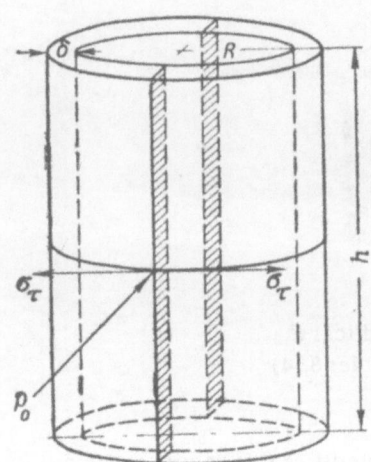

Fig. 25. The case of a thin-walled cylindrical shell.

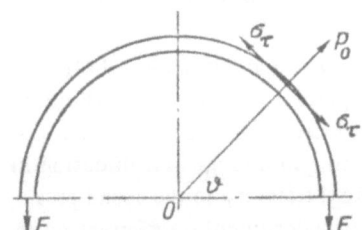

Fig. 26. Calculation of the force F acting on the semicylinder.

equal in view of the spherical symmetry of the problem. Knowing the total external force F acting on the hemisphere in the direction perpendicular to the equatorial cross section, we can determine σ_τ and σ_ψ.

Let the thickness of the shell be $\delta (\delta \ll R)$. The area of the equatorial cross section is then $S = 2\pi R \delta$. The total force acting in a direction normal to this cross section is given by the integral

$$F = \int_0^{\pi/2} p\, dS \quad \text{or} \quad F = 2\pi p_0 R^2 \int_0^{\pi/2} \sin \vartheta \cos \vartheta\, d\vartheta.$$

The notation employed is shown in Fig. 24. Integration gives

$$F = p_0 \pi R^2,$$

and consequently the mechanical stress σ_τ is given by $\sigma_\tau = p_0(R/2\delta)$. Similarly,

$$\sigma_\psi = p_0 \frac{R}{2\delta} = \sigma_\tau = \sigma.$$

We now assume that the small element of the shell shown in Fig. 23 consists of piezoelectric material polarized in the direction of the radius of the sphere. Then the emf developed by the piezoelectric element due to the mechanical stress σ_τ will be $E_\tau = \delta g_{31} \sigma_\tau$ and consequently the emf created by the stress σ will be $E_\psi = \delta g_{31} \sigma_\psi$. The total emf developed by the piezoelectric element will be

$$E = E_\tau + E_\psi,$$

or assuming that $\sigma_\tau = \sigma_\psi = \sigma$ and $g_{31} = g_{32} = g$,

$$E = 2\delta\sigma g.$$

Substituting in this expression the previously obtained expression for σ, we finally get

$$E = R\, p_0 g = \frac{R}{\delta} \delta p_0 g,$$

and, consequently, the coefficient of mechanical transformation for a thin-walled shell is

$$K = \frac{R}{\delta}.$$

Let us now consider the case of a thin-walled cylindrical shell (Fig. 25) whose external surface is subjected to a given excess pressure p_0. We indicate the external radius by R, the axial length by h, and the wall thickness by δ with $\delta \ll R$. We assume that the end faces of the cylinder are shielded from the effects of the sound pressure. Thus, the cylinder can freely deform in the axial direction and the mechanical stresses at any transverse cross section of the cylindrical still will be equal to zero ($\sigma_h = 0$). Tangential mechanical stresses σ_τ will arise only in diametral cross sections of the shell, i.e., in the direction perpendicular to the generator of the cylinder.

We now dissect the cylinder along the plane including the axis and determine the stress acting in the cross section thus formed. The area of this cross section is $S = 2h\delta$. The force F acting on this section is expressed by the integral

$$F = h \int_0^\pi p\, dS \quad \text{or} \quad F = h p_0 R \int_0^\pi \sin \vartheta\, d\vartheta,$$

which employs the notation given in Fig. 26. Integration gives

$$F = 2hR p_0.$$

Consequently, the mechanical stress σ_T is

$$\sigma_\tau = p_0 \frac{R}{\delta}.$$

We now cause a piezoelectric element to be inserted in the shell. The element is polarized as shown in Fig. 27. The emf developed by such an element is given by

$$E = E_\tau = \delta g_{31} \delta_\tau = p_0 \delta \frac{R}{\delta} g,$$

which leads to the following simple expression for the coefficient of mechanical transformation for the case of a thin-walled cylindrical shell:

$$K = \frac{R}{\delta}.$$

Thus, the coefficients of mechanical transformation for thin-walled spheres and cylinders are identical for the same values of the radius R and shell thickness δ.

At sufficiently high values of the ratio R/δ it is possible to obtain high transformation coefficients and consequently to increase the sensitivity of the shell-type sound receiver incorporating a piezoelectric element by some tens or hundreds of times in comparison with the sensitivity of the piezoelectric element alone.

In what has been said, we have assumed specific orientations of the piezoelectric elements in relation to the total structure of the shell and made certain assumptions concerning the properties of the piezoelectric element in order to simplify the explanation. It is clear that the coefficient of mechanical transformation can

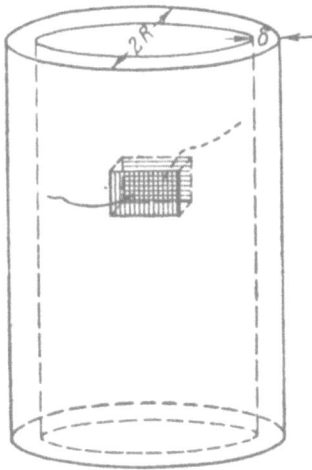

Fig. 27. Disposition of the electrodes with radial polarization of
a ceramic cylinder.

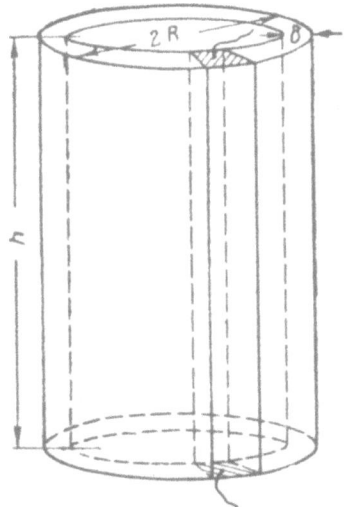

Fig. 28. A simple method of electrode arrangement with longitudinal
polarization of a ceramic cylinder.

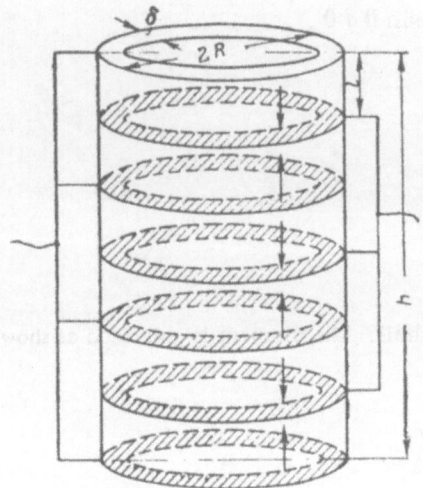

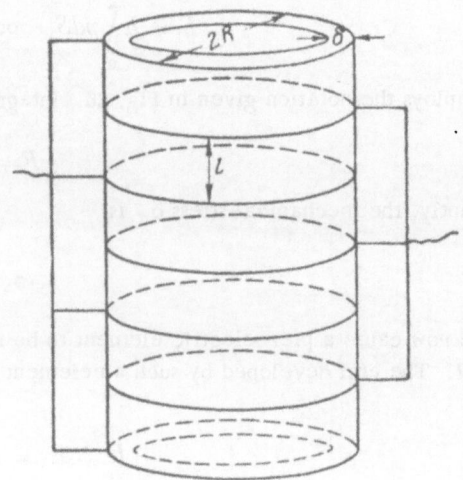

Fig. 29. Electrode arrangement using a sectioned cylinder for longitudinal polarization.

Fig. 30. Electrodes located on the external curved surface with longitudinal polarization of a thin continuous cylinder.

be calculated for any orientation of the piezoelectric element and for any piezoelectric parameters; in such cases it is only necessary to take account of the corresponding relationships between the piezoelectric coefficients. Naturally, the conclusions reached are valid only when the dimensions of the piezoelectric element are small in comparison with the shell dimensions and when the mechanical parameters of the element differ only slightly from those of the shell. If these conditions are not satisfied the problem becomes more complex. However, as will be demonstrated below, in the case of piezoceramic sound receivers it is preferable to construct the shell and the piezoelectric element of the same material, in which case no correction is required for differences in the parameters of the corresponding materials.

Let us consider the principal means by which it is possible to use continuous ceramic shells as sound receivers. These methods arise firstly out of the possibility of using individual regions of a shell or the complete shell as the piezoelectric transducer element and secondly from the possibility of using different types of polarization of the piezoceramic in those regions of the shell selected as transducer elements.

The simplest method employing a cylindrical shell is shown in Fig. 27. Electrodes of equal area are deposited opposite one another on the external and internal surfaces of the shell. Naturally, in this method it is possible to use only radial polarization of the piezoceramic. The region of the shell sandwiched between the electrodes acts as the transducer; the remainder of the shell acts as a mechanical stress transformer. On the basis of our previous results we at once obtain an expression for the sensitivity of the given sound receiver in the form

$$\frac{E}{p_0} = K\delta g_{31} = Rg_{31}.$$

We see that the sensitivity in this case does not depend on the area occupied by the electrodes; this factor affects only the capacity of the transducer. The sensitivity can only be increased by increasing the radius of the cylinder. Since in most cases it is advantageous to increase the capacity of the transducer, the electrodes should cover the entire inner and outer surface of the shell.

Figure 28 shows another simple method of using a cylinder shell. In this case the electrodes are deposited opposite one another on the end faces of the shell. In this case the sensitivity is

$$\frac{E}{p_0} = Khg_{31} = \frac{R}{\delta}hg_{31}.$$

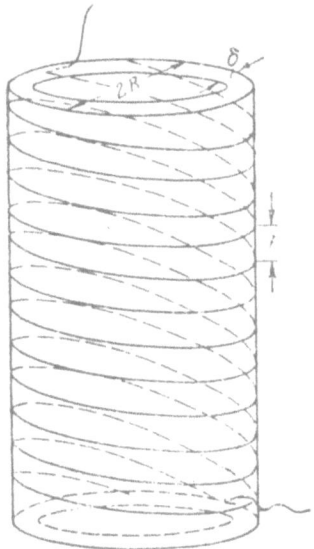

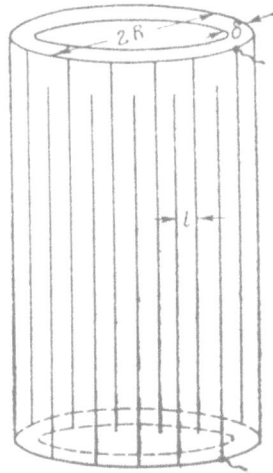

Fig. 31. Electrode system in the form of a double-start spiral for longitudinal polarization of a thin-walled continuous cylinder.

Fig. 32. Electrodes deposited on the curved surface of a thin-walled tangentially polarized ceramic cylinder.

We see that in this case the sensitivity can be increased either by decreasing the shell thickness or increasing the axial length h of the cylinder. However, this would unavoidably decrease the capacity of the transducer. This case has been referred to in the literature as the case of "longitudinal" polarization of a cylindrical piezoelectric transducer.

Naturally, the capacity can be increased by making the electrodes cover the entire end surfaces of the cylindrical shell. However, in most cases the capacity would still be too small. Of course, if we disregard the sensitivity, it is possible to increase the capacity but for this it is necessary to subdivide the cylindrical shell into a number of sections of small height, e.g., by fusing together a number of short cylindrical elements with end electrodes. In this case, the polarization in neighboring rings should be in opposite directions so as to obtain parallel connection of the sections of the transducer. If the number of sections is n, the sensitivity is

$$\frac{E}{p_0} = \frac{R}{\delta} \, l g_{31} = \frac{R}{\delta} \, \frac{h}{n} \, g_{31} \; ,$$

where l is the axial length of the section. Thus, the sensitivity of a sectioned transducer of the given type is decreased by n times in comparison with the sensitivity of a continuous cylinder with end surface electrodes (for the same overall axial length), but the capacity is increased n times. The fusing together of the sections is easily achieved only in the case of relatively large shell thicknesses. However, in the case of thin-wall shells it is not necessary to section the cylinder as shown in Fig. 29. The high piezoelectric constant of piezoceramics enables the realization of longitudinal polarization with the electrode system shown in Fig. 30 whereby ring-shaped electrodes are deposited only on the external surface of the cylindrical shell with a spacing l.

Since in this case the shell thickness is much smaller than the distance between the electrodes and the dielectric constant of the ceramic is three orders greater than that of the surrounding medium, the polarizing field in the region between the electrodes is sufficiently uniform. However, it is still necessary to use a voltage 30 to 40% higher than the usual value.

Perhaps the most convenient electrode form for use in the given case is the double-start spiral configuration shown in Fig. 31 [194] which avoids the soldering of leads to electrodes.

TABLE 9

Method of polarization	Sensitivity, μV/bar	Sensitivity ratio	Capacity	Capacity ratio
Radial	$Rg_{31} = 4.7R$	1	$\dfrac{\varepsilon Rh}{2\delta}$	1
Longitudinal	$Rg_{31}\dfrac{l}{\delta} = 4.7R\dfrac{l}{\delta}$	$\dfrac{l}{\delta}$	$\dfrac{\varepsilon R\delta}{2l}$	$\left(\dfrac{\delta}{l}\right)^2\dfrac{l}{h}$
Tangential	$\dfrac{2\pi R^2}{\delta n}g_{33} = 71\dfrac{R^2}{\delta n}$	$\left(\dfrac{2\pi R}{n\,\delta}\right)\dfrac{g_{33}}{g_{31}}$	$\dfrac{\varepsilon h n^2\delta}{8\pi^2 R}$	$\left(\dfrac{n\delta}{2\pi R}\right)^2$

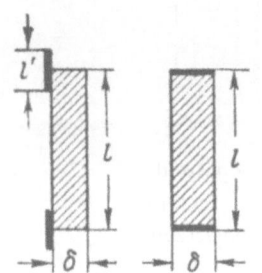

Fig. 33. Equivalent elec-
trode forms.

However, the method of surface deposition of the electrodes enables the sensitivity of a cylindrical sound receiver to be considerably increased by changing from radial or longitudinal polarization to tangential polarization. The corresponding electrode system is shown in Fig. 32. If the distance l between the electrodes is equal to $2\pi R/n$ where n is the number of sections around the circumference of the cylinder, the sensitivity is given by

$$\frac{E}{p_0} = Klg_{33} = \frac{2\pi R^2}{\delta n}g_{33}\,.$$

We see that in this case the sensitivity is formally expressed in the same way as for longitudinal polarization except that the constant g_{33} is used instead of g_{31}. Thus, for otherwise identical conditions the sensitivity increases in proportion to g_{33}/g_{31}.

The sensitivity and other parameters of thin-walled cylindrical sound receivers are compared in Table 9 for different electrode systems and polarizations. The table gives formulae for determining the sensitivity and capacity and expressions which characterize the ratios of these values to the corresponding values for sound receivers with radial polarization and electrodes which occupy the whole of the internal and external surfaces. Numerical coefficients are given for ceramics with the mean parameters cited at the end of Chapter I. The table employs the following notation: R, external radius of the cylinder; h, axial length of the cylinder; δ, wall thickness of the ceramic shell; l, distance between electrodes; n, number of sections along the cylinder length or around the circumference of the cylinder. The expressions with numerical coefficients give the sensitivity directly in μV/bar if the geometrical dimensions are in centimeters.

In determining the capacity for surface deposition of the electrodes we assume that the surface-deposited electrode is equivalent to an electrode deposited at the end surface of the section if the sections were fused together (Fig. 33). This assumption is valid for $l' \ll l \approx \delta$.

We shall now consider the possible methods of using a thin-walled spherical shell. In this case we must choose between radial and tangential polarization. On the basis of the same considerations as for the case of radial polarization, it is advantageous to use continuous electrodes deposited on the internal and external surfaces of the spherical shell. In this case, the sensitivity of a spherical sound receiver is

$$\frac{E}{p_0} = K\delta g_{31} = Rg_{31}. \tag{4}$$

If for any reason it appears necessary to reduce the capacity of the transducer, it is possible to deposit electrodes opposite each other which only cover part of the internal and external surfaces. In this case the sensitivity remains unaltered.

It is of course possible to subdivide the electrodes so as to form a series connection of electrode pairs as shown in Fig. 34. By this means it is possible to increase the sensitivity with a corresponding decrease in

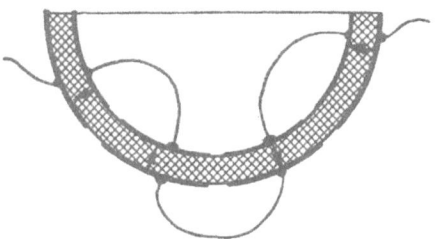

Fig. 34. Series connection of individual sections of a radially polarized, spherical ceramic shell.

Fig. 35. A photograph showing the two halves of a tangentially polarized spherical piezoelectric element constructed from barium titanate ceramic.

capacity. The configuration of the sections at the surface of the sphere should be such that the sections are equal in area. The regions of the shell between the electrodes should be wider than the thickness of the shell.

Naturally, tangential polarization cannot be achieved over the shell as a whole; the electrodes must be distributed in some manner over the external surface. One possible method is shown in Fig. 35. Assuming that the tangentially polarized regions occupy most of the surface of the sphere, the sensitivity of a tangentially polarized spherical sound receiver is given by the formula

$$\frac{E}{p_0} = Klg_{33} \ .$$

It is obvious that for otherwise equal conditions, tangential polarization is more advantageous since the expression for the sensitivity contains the piezoelectric coefficient d_{33}.

Thus, we reach the conclusion that maximum sensitivity is achieved by the use of tangentially polarized cylindrical or spherical shells. In this case, in addition to increase in sensitivity arising from mechanical transformation of the stresses (which occurs for any type of polarization of thin-walled shells) it is possible to make use of the longitudinal piezoelectric coefficient d_{33} which is more than twice as great as the transverse piezoelectric coefficient $d_{31} = d_{32}$. In addition, for a given transformation coefficient R/δ the use of tangential polarization permits a greater distance l between the electrodes. In other words, the use of tangential polarization increases the sensitivity due to increase in the ratio R/δ, increase in l, and utilization of the piezoelectric coefficient d_{33}. Longitudinal polarization results in increase in the ratio R/δ and increase in l only. Thus, the sensitivity of a longitudinally polarized cylindrical sound receiver is less than that of a tangentially polarized sound receiver with the same values of R, l, and δ. For given shell dimensions radial polarization of spherical and cylindrical shells gives the lowest sensitivity but at the same time gives maximum capacity.

In the case of a radially polarized spherical shell the sensitivity can only be increased by increase in the external radius since an increase in shell thickness δ leads to a decrease in the coefficient of mechanical transformation.

Thus, the use of thin-walled ceramic shells as sensitive elements in sound receivers enables the parameters of the latter to be varied, e.g., it is possible to design a receiver of low capacity and high sensitivity, or conversely. The type of receiver depends on the purpose for which it is to be used and the conditions under which it is used (e.g., on the minimum permissible distance from the first cascade amplifier, etc.). Both types of receiver are of practical interest.

Variation of the parameters of a sound receiver by the use of various electrode systems at the surface of a ceramic shell is to some extent equivalent to series or parallel connection of individual elementary piezoelectric elements. In this way we solve the problem of matching the electrical impedance of the sound receiver to the input stage of the amplifier. In certain special cases, e.g., in the case of low-frequency sound receivers, matching can be achieved by the use of an input transformer. However, in comparison with simple plane- or assembly-type elements, the use of closed thin-walled shells as piezoelectric elements possesses the great advantage of the utilization of mechanical stress transformation in the sound receiver material. Increase in mechanical stress in the material leads to increased sensitivity of the sound receiver.

TABLE 10

Form of receiver surface and type of polarization	Ratio of the reactive electric power to the specific acoustic power
Plane element using the piezoelectric coefficient d_{33}	$V\omega\rho c\; \dfrac{4\pi d_{33}^2}{\varepsilon}$
Plane element using the piezoelectric coefficient $d_{32}=d_{31}$	$V\omega\rho c\; \dfrac{4\pi d_{32}^2}{\varepsilon}$
Radially polarized sphere	$V\omega\rho c\; \dfrac{4\pi d_{32}^2}{\varepsilon}\left(\dfrac{R}{\delta}\right)^2$
Radially polarized cylinder	$V\omega\rho c\; \dfrac{4\pi d_{32}^2}{\varepsilon}\left(\dfrac{R}{\delta}\right)^2$
Longitudinally polarized cylinder	$V\omega\rho c\; \dfrac{4\pi d_{32}^2}{\varepsilon}\left(\dfrac{R}{\delta}\right)^2$
Tangentially polarized cylinder	$V\omega\rho c\; \dfrac{4\pi d_{33}^2}{\varepsilon}\left(\dfrac{R}{\delta}\right)^2$

The importance of mechanical transformation can be evaluated by employing as parameter the ratio of the reactive electrical power developed by a piezoelectric element of a given volume of piezoelectric material when placed in the field of a plane acoustic wave of known acoustic power.

If we ignore the internal losses in the ceramic, the results of such an evaluation can be expressed by means of the formulae given in Table 10. This table uses the following notation: V, volume of the piezoelectric material; ω, angular frequency; ρc, characteristic impedance of the medium in which the sound receiver is located; ε, dielectric constant of the material; R, radius of the shell; δ, wall thickness of the shell. Table 10 shows that for the same volume of material the use of mechanical stress transformation results in an increase in the reactive electric power proportional to the square of the ratio of the radius of the cylindrical or spherical shell to the wall thickness of the shell.

The data for cylinders given in Table 10 relate to the case when the end surfaces are shielded from the effect of the sound pressure. This is the optimum method for use with tangentially or longitudinally polarized cylindrical elements.

Thus, calculation of the static sensitivity of barium titanate ceramic sound receivers shows that the use of thin-walled spherical or cylindrical shells results in higher sensitivity compared with receivers of simpler construction using rods or assembly-type piezoelectric elements [195-197]. This arises from the more rational use of the piezoelectric material in spherical or cylindrical elements.

However, in the case of nondirectional measurement sound receivers special demands are sometimes made with respect to mechanical strength (e.g., when investigating detonation effects or sound pressures in deep water, etc.). In such cases, some loss in sensitivity must be accepted and the wall thickness of the shell increased. Thus, it is necessary to determine the relationship between the sensitivity of a cylindrical or spherical sound receiver and its dimensions for the case of relatively high wall thickness.

3. Static Sensitivity of a Radially Polarized Spherical Piezoelectric Element with a Finite Wall Thickness

Let a and b be the internal and external radii of a spherical sound receiver. The external surface of the spherical shell is subjected to an excess pressure $p_b = p_0$; within the shell the excess pressure $p_a = 0$. We shall consider a volume element of the piezoelectric material bounded by two spherical surfaces with radii R and

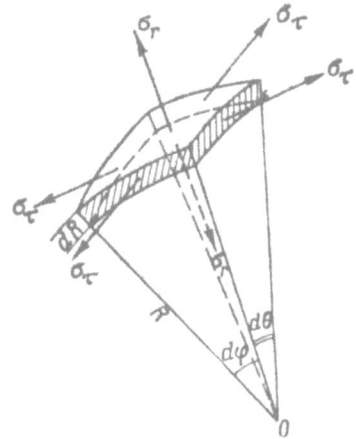

Fig. 36. Volume element bounded by two spherical surfaces.

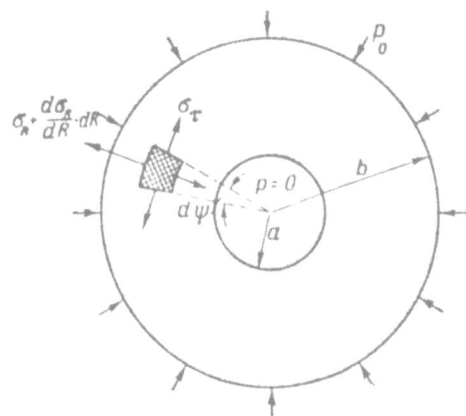

Fig. 37. Method of calculating the strain σ_τ

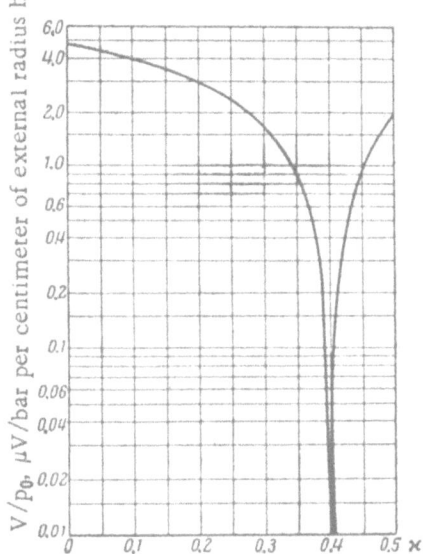

Fig. 38. Calculation of the static sensitivity of radially polarized spherical piezoelectric elements of barium titanate ceramic with different wall thicknesses.

R + dR, two mutually perpendicular planes passing through the center of the sphere, and two planes which also pass through the center of the sphere and located at angles of $d\theta$ and $d\varphi$, respectively, with respect to the first two planes (Fig. 36).

In view of the symmetry of the shell and also the symmetry of the external forces acting on the shell, the intensity of the electric field created by the deformation of the elementary volume is directed radially. The field intensity E depends on the mechanical stresses σ_R, σ_T and σ_T corresponding to the radial and two mutually perpendicular directions. We can write

$$E = E_R + E_\theta + E_\varphi,$$

where $E_R = g_{33}\sigma_R$, $E_\theta = g_{31}\sigma_T$ and $E_\varphi = g_{31}\sigma_T$.

Thus,

$$E = g_{33}\sigma_R + 2g_{31}\sigma_\tau.$$

It is clear that E depends on R and that this dependence becomes manifest in the dependence of the mechanical stresses σ_R and σ_T on R. The potential difference set up between the external and internal surfaces of the spherical element is given by the integral

$$V = \int_a^b (g_{33}\sigma_R + 2g_{31}\sigma_\tau)dR. \tag{5}$$

The general expression for the radial stress σ_R in the spherical shell takes the form [198]

$$\sigma_R = \frac{C}{R^3} + D,$$

45

where C and D are arbitrary constants determined by the boundary conditions $\sigma_{R=b} = -p_0$ and $\sigma_{R=a} = 0$. Hence

$$C = \frac{a^3 b^3}{b^3 - a^3} p_0, \qquad D = -\frac{b^3}{b^3 - a^3} p_0,$$

and, consequently,

$$\sigma_R = \frac{b^3}{b^3 - a^3} \frac{a^3 - R^3}{R^3} p_0. \tag{6}$$

The expression for the stress σ_τ is easily obtained from the equilibrium conditions of an element cut from a spherical shell and bounded by two concentric surfaces of radii R and R + ΔR and a circular cone with its apex at the center of the sphere and a small apex angle dφ (Fig. 37). In this case, we get

$$\sigma_\tau = \frac{d\sigma_R}{dR} \frac{R}{2} + \sigma_R ,$$

where $d\sigma_R/dR$ and σ_R are determined from (6). The final expression for the stress σ_τ takes the form [199]

$$\sigma_\tau = \frac{b^3 (2R^3 + a^3)}{2R^3 (a^3 - b^3)} \cdot p_0 .$$

Substituting the values of σ_τ and σ_R in (4) gives the expression for the sensitivity of a radially polarized ceramic sphere in the form

$$\frac{V}{p_0} = \int_a^b \left(g_{33} \frac{b^3}{b^3 - a^3} \frac{a^3 - R^3}{R^3} + g_{31} \frac{b^3}{R^3} \frac{2R^3 + a^3}{a^3 - b^3} \right) dR$$

or

$$\frac{V}{p_0} = \frac{b^3}{b^3 - a^3} \left[g_{33} a^3 \int_a^b \frac{1}{R^3} dR - g_{33} \int_a^b dR - 2g_{31} \int_a^b dR - a g^3_{31} \int_a^b \frac{dR}{R^3} \right] .$$

After integration we get

$$\frac{V}{p_0} = \frac{b^3}{b^3 - a^3} \left[(g_{33} - g_{31}) - \frac{a(b^3 - a^3)}{2b^3} - (b - a)(g_{33} + 2g_{31}) \right] =$$

$$= \frac{b}{b^3 + a e + a^3} \left[(g_{33} - g_{31}) \frac{a(b + a)}{2} - b^3 (g_{33} + 2g_{31}) \right] . \tag{7}$$

If the ratio of the internal and external radii of the sphere is indicated by a/b = η, then after some algebraic rearrangement formula (7) takes the simpler form

$$\frac{V}{p_0} = \frac{b}{\eta^3 + \eta + 1} \left[\frac{\eta(1 + \eta)}{2} (g_{33} - g_{31}) - (g_{33} + 2g_{31}) \right],$$

or

$$\frac{V}{p_0} = \frac{b}{\eta^2 + \eta + 1} \left[\frac{\eta^2 + \eta - 2}{2} g_{33} - \frac{\eta^2 + \eta + 4}{2} g_{31} \right] .$$

The formula for the receiver sensitivity V/p_0 can be conveniently analyzed by representing the sensitivity as a function of the ratio of wall thickness to external diameter of the shell

$$\varkappa = \frac{b - a}{2b} ;$$

TABLE 11

External diameter of the spherical shell D, cm	Wall thickness of the spherical shell δ, cm	Working frequency range, kc	Capacity, μμF	Sensitivity, μV/bar	
				Theoretical	Experimental
1.8	0.05	0-120	20,000	4.05	3.0
2.2	0.40	0-115	1,700	2.20	2.0
2.1	0.35	0-120	2,300	2.20	2.0
2.0	0.25	0-120	3,900	3.00	2.9
1.9	0.15	0-120	6,300	3.42	3.0

then

$$\frac{V}{p_0} = b \frac{1}{3 - 6\varkappa + 4\varkappa^2} \left[(2\varkappa^2 - 3\varkappa + 1)(g_{33} - g_{31}) - (g_{33} + 2g_{31}) \right] =$$
$$= b \frac{1}{3 - 6\varkappa + 4\varkappa^2} \left[g_{33}(2\varkappa^2 - 3\varkappa) - g_{31}(2\varkappa^2 - 3\varkappa + 3) \right]. \tag{8}$$

We see from (7) that the sensitivity of a radially polarized ceramic sphere is proportional to the external radius of the sphere. Therefore, it is convenient to use a graph constructed for the case b = 1 cm (Fig. 38). This graph was constructed for specific values of g_{33} and g_{31} used by us in all the calculations of the sensitivities of ceramic transducers (see Chapter I).

It is clear from (7) and from the graph that for a given value of $\varkappa$ the sensitivity becomes zero. This value of $\varkappa$ is easily obtained by equating to zero the expression in square brackets in (7):

$$g_{33}(2\varkappa^2 - 3\varkappa) = g_{31}(2\varkappa^2 - 3\varkappa + 3),$$

from which we find that for ceramics with $g_{33}/g_{31} = 2.4$, the sensitivity of radially polarized spherical sound receivers is zero for $\varkappa = 0.402$. For $\varkappa = 0$, i.e., for a wall thickness negligibly small in comparison with the shell diameter, we get

$$\frac{V}{p_0}\bigg|_{\varkappa = 0} = -bg_{31}.$$

This expression is in agreement with the formula obtained earlier for a thin-walled shell [formula (4)].[*] For $\varkappa = \frac{1}{2}$, i.e., for a continuous radially-polarized ceramic sphere, we get

$$\frac{V}{p_0}\bigg|_{\varkappa = \frac{1}{2}} = -b(2g_{31} + g_{33}).$$

It is easily seen that the expression in brackets is nothing other than the hydrostatic coefficient. In order to show the extent to which the theoretical assumptions concerning the effect of shell thickness affect practical results, the parameters of a number of experimental spherical sound receivers with external diameters of approximately 2 cm and various wall thicknesses are given in Table 11. This table shows that the static sensitivity calculated from (8) is in good agreement with the experimental data.

4. Static Sensitivity of a Cylindrical Piezoelectric Element of Finite Wall Thickness for Various Types of Polarization

Langevin [200] obtained expressions giving the sensitivity of cylindrical piezoelectric elements of finite wall thickness for various types of polarization and different conditions at the end faces of the cylinders. Since it is advantageous to use thick-walled cylindrical piezoelectric elements in sound receivers destined for use at high hydrostatic pressures, we have considered it desirable to reproduce the principal results of this work.

---

[*]Formula (4) was obtained without regard to sign.

TABLE 12

Polarization	Conditions at the end faces
Radial	End faces shielded from the sound pressure
	Ring-shaped end faces exposed to the sound pressure
	End faces covered by caps which receive the sound pressure. The diameter of the caps was equal to the external diameter of the piezoelectric element.
Longitudinal (along z axis)	End faces shielded from the sound pressure
	Ring-shaped end faces exposed to the sound pressure
	End faces covered by caps which receive the sound pressure. The diameter of the caps was equal to the external diameter of the piezoelectric element.
Tangential	End faces shielded from the sound pressure
	Ring-shaped end faces exposed to the sound pressure
	End faces covered by caps which receive the sound pressure. The diameter of the caps was equal to the external diameter of the piezoelectric element.

Langevin considered different types of polarization and various conditions at the end faces (Table 12).

For convenience in comparing the data of Langevin with our own expressions for the sensitivity of thin-walled cylindrical piezoelectric elements we have rewritten the Langevin formulae in our notation. In the case of cylindrical symmetry of the piezoelectric element we naturally select the coordinates R, θ, and z in considering the volume element of the piezoelectric material (Fig. 39). The mechanical stresses in the shell are correspondingly indicated by σ_R, σ_θ, and σ_z.

In the case of radial polarization, i.e., in the case of electrodes deposited on the external and internal surfaces of the cylindrical shell, the potential difference between the electrodes is given by the integral

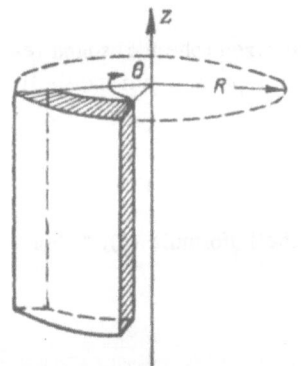

Fig. 39. Volume element used in calculating the static sensitivity of a cylindrical piezoelectric element.

$$V = \int_a^b [g_{33}\sigma_R + g_{31}(\sigma_\theta + \sigma_R)]\, dR,$$

where b and a are the external and internal radii, respectively, of the cylindrical shell.

In the case of longitudinal polarization, i.e., with electrodes located at the ring-shaped end surfaces of the piezoelectric element, we get

$$V = \frac{1}{\pi(b^2 - a^2)} \int_0^{2\pi} \int_a^b g_{33}\sigma_z + g_{32}(\sigma_R + \sigma_\theta)\, Rd\, Rd\,\theta,$$

and finally in the case of tangential polarization with electrodes subdividing the cylindrical element into n sections, the potential difference is given by

$$V = \frac{2\pi}{nl \ln \frac{b}{a}} \int\limits_0^l \int\limits_a^b [g_{33}\sigma_\theta + g_{32}(\sigma_R + \sigma_z)]\, dR\, dz.$$

The mechanical stresses σ_R, σ_θ, and σ_z in these expressions depend on the conditions at the end faces of the piezoelectric element. When the end faces are shielded from the sound pressure, we get

$$\sigma_R = \frac{a^2 b^2 p_0}{b^2 - a^2}\left(\frac{1}{R^2} - \frac{1}{a^2}\right),$$

$$\sigma_\theta = \frac{a^2 b^2 p_0}{b^2 - a^2}\left(-\frac{1}{R^2} - \frac{1}{a^2}\right), \quad \sigma_z = 0.$$

With the sound pressure acting on the ring-shaped end faces, we get

$$\sigma_R = \frac{a^2 b^2 p_0}{b^2 - a^2}\left(\frac{1}{R^2} - \frac{1}{a^2}\right),$$

$$\sigma_\theta = \frac{a^2 b^2 p_0}{b^2 - a^2}\left(-\frac{1}{R^2} - \frac{1}{a^2}\right),$$

$$\sigma_z = -p_0,$$

and, finally in the presence of caps taking the sound pressure, we get

$$\sigma_R = \frac{a^2 b^2 p_0}{b^2 - a^2}\left(\frac{1}{R^2} - \frac{1}{a^2}\right),$$

$$\sigma_\theta = \frac{a^2 b^2 p_0}{b^2 - a^2}\left(-\frac{1}{R^2} - \frac{1}{a^2}\right),$$

$$\sigma_z = -\frac{b^2}{b^2 - a^2}\, p_0.$$

Here p_0 everywhere indicates the external excess pressure. The excess pressure inside the cylinder is assumed zero.

Substituting the expressions for the stresses in the corresponding integrals giving the potential difference developed by the piezoelectric element we get the expression for the static sensitivity of a cylindrical piezoelectric element (Table 13). The coefficient A in Table 13 is defined as

$$A = \frac{2\pi b}{n \ln \frac{1}{1 - 2\varkappa}},$$

where l is the axial length of the cylinder.

The formulae given in Table 13 were used to construct graphs for determining the sensitivity of a ceramic piezoelectric element with given piezoelectric coefficients. Corresponding graphs given in Langevin's paper relate to a piezoceramic for which the coefficients g_{33} and g_{31} exceed by only 10% the values employed in our calculations.

Figure 40a shows the graph for a radially polarized cylindrical piezoelectric element. The corresponding formulae show that the sensitivity is proportional to the external radius of the cylinder; therefore the graph was constructed for b = 1.

The figure contains three curves: for end faces shielded from the sound pressure (curve I), for ring-shaped end surfaces exposed to the sound pressure (curve II), and for capped end surfaces (curve III). It is easily seen that capped end surfaces give increased sensitivity; in the case of a thin-walled shell ($\varkappa = 0$) this increase amounts to about 50%. As in the case of a spherical shell, the sensitivity of a cylindrical element drops to zero at specific values of $\varkappa$ dependent finally on the conditions at the end surfaces.

TABLE 13

Polarization	Conditions at the end faces		
	End faces shielded from the sound pressure	Ring-shaped end faces exposed to the sound pressure	End faces covered by caps taking the sound pressure
Radial	$b\left[\dfrac{\varkappa}{1-\varkappa}g_{33}+g_{31}\right]$	$b\left[\dfrac{\varkappa}{1-\varkappa}g_{33}+(1-2\varkappa)g_{31}\right]$	$b\left[\dfrac{\varkappa}{1-\varkappa}g_{33}+\dfrac{3-2\varkappa}{2-2\varkappa}g_{31}\right]$
Longitudinal	$l\left[\dfrac{g_{31}}{2(\varkappa-\varkappa^2)}\right]$	$l\left[\dfrac{g_{31}}{2(\varkappa-\varkappa^2)}+g_{33}\right]$	$l\left[\dfrac{2g_{31}+g_{33}}{4(\varkappa-\varkappa^2)}\right]$
Tangential	$A\left[g_{31}\dfrac{\varkappa}{1-\varkappa}+g_{33}\right]$	$A\left[g_{31}\dfrac{\varkappa(3-2\varkappa)}{1-\varkappa}\right]+g_{33}$	$A\left[g_{31}\dfrac{1-2\varkappa}{2(1-\varkappa)}+g_{33}\right]$

Figure 40b shows the graph for longitudinal polarization. Since in this case the sensitivity is proportional to the axial length of the cylinder, the graph is constructed for $l=1$. As in the case of Fig. 40a three curves are given; for end faces shielded from the sound pressure (curve I), ring-shaped end faces exposed to the sound pressure (curve II), end faces covered by caps of diameter 2b (curve III). A zero sensitivity is encountered only for curve II.

Finally, Fig. 40c shows the curves for tangential polarization for n = 2 and b = 1. The notation of the curves is the same as for Figs. 40a and 40b. In this case the sensitivity naturally drops to zero for $\varkappa = 0.5$ since for this value of $\varkappa$ the coefficient A becomes zero.

It is of interest to examine the question: for which wall thickness-diameter ratios $\varkappa$ is it possible to use the formulae given in Section 2 and under what conditions does it become necessary to use the more accurate calculations?

A comparison [196] shows that the approximation formulae can be used for measurement sound receivers, for use in tanks, shallow water, and in air when $\varkappa$ cannot be higher than 0.05 to 0.06.

Figure 41 shows the dependence of the static sensitivity of radially-polarized cylindrical and spherical sound receivers on $\varkappa$ for low values of this parameter. The broken horizontal straight lines give the sensitivity calculated on the basis of the approximation formulae; the continuous lines correspond to the accurate formulae. Curve 1 relates to a sphere; curve 2 relates to a cylinder with its end surface not shielded against the sound pressure; curve 3 is based on the approximation formulae corresponding to these two cases; curve 4 corresponds to the case of a cylinder with rigid end caps of diameter 2b; curve 5 corresponds to the same case but based on the approximation formulae.

From Fig. 41 we see that for $\varkappa = 0.05$ the discrepancy between the results given by the approximation and accurate formulae amount to 2.1 db for a sphere, 1.2 db for a cylinder with unshielded end faces, and 0.73 db for a cylinder with end caps. The same comparison is made in Fig. 42 for a longitudinally polarized cylinder with unshielded end faces.* In this case, the discrepancy between the results given by the accurate (continuous line curve) and approximation (broken line curve) formulae amounts to 0.46 db for $\varkappa = 0.05$.

Figure 43 compares the results obtained by the approximation and accurate calculations for a tangentially polarized cylinder. In this case, for $\varkappa = 0.05$ the discrepancy amounts to 1.16 db. Thus for $\varkappa \ll 0.05$ it is sufficient to use the approximation formulae in calculating the sensitivity of ceramic sound receivers.

It is interesting to compare the sensitivity of thin-walled piezoelectric elements of various types with solid ceramic forms of equivalent external dimensions. In these limit cases the sensitivity formulae become very simple. Such a comparison is made in Fig. 44 in which the piezoelectric elements arranged in each row

---

* The caption to Fig. 42 states that the end faces are shielded from the acoustic pressure.— Translator's Note

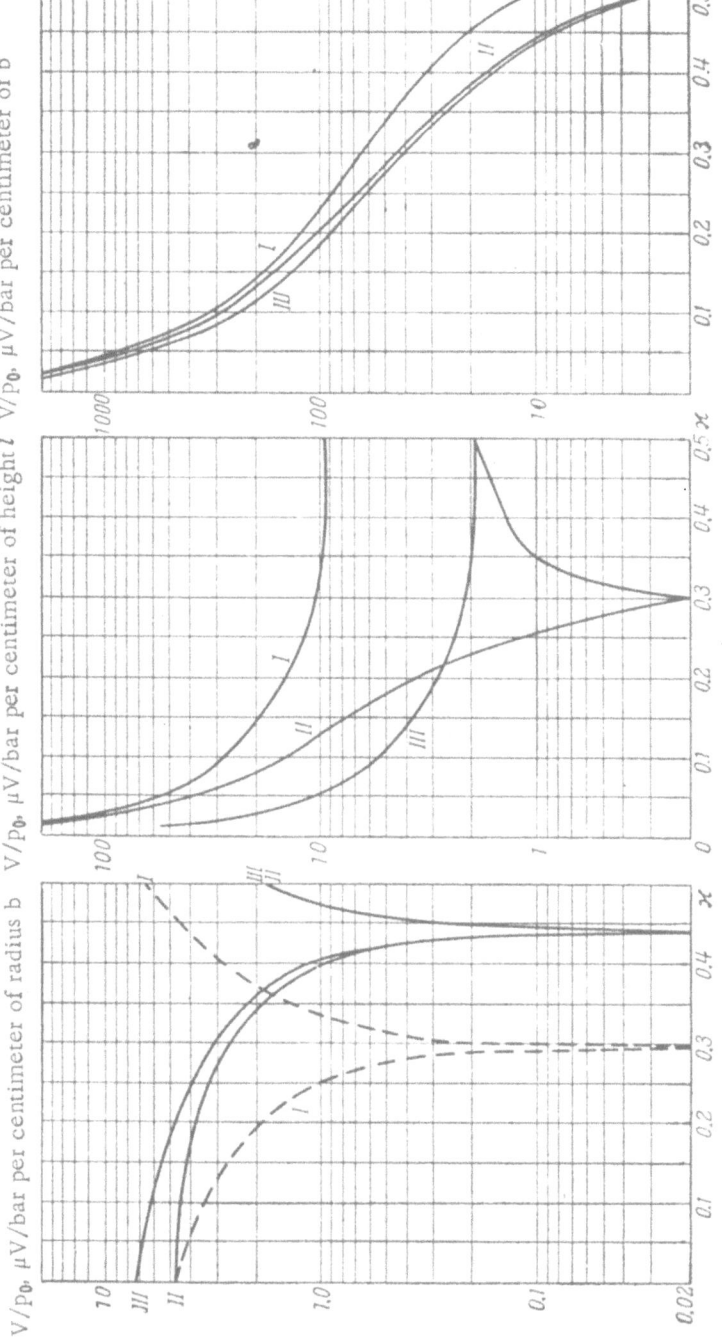

Fig. 40. Sensitivity of barium titanate ceramic cylinders for different ratios of wall thickness to external diameter $\varkappa$, $b = 1$ cm. a) Radial polarization; b) longitudinal polarization; c) tangential polarization.

V/p₀, µV/bar per centimeter of radius b

V/p_0, µV/bar per centimeter of radius b

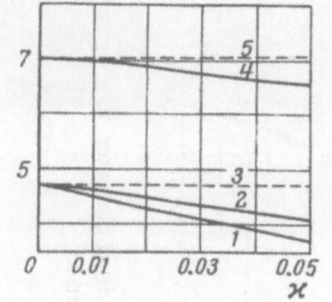

Fig. 41. Calculation of the static
sensitivity of radially· polarized
spherical and cylindrical shells
of barium titanate ceramic.

V/p_0, µV/bar per centimeter of l

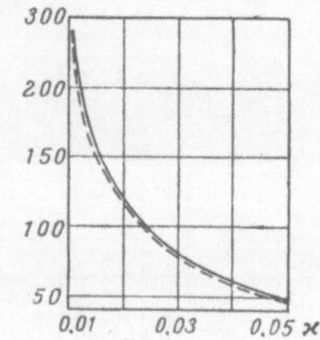

Fig. 42. Calculation of the static
sensitivity of a longitudinally pol-
larized cylinder. End faces of the
cylinder shielded from the acoustic
pressure.

V/p_0, µV/bar per centimeter of radius b

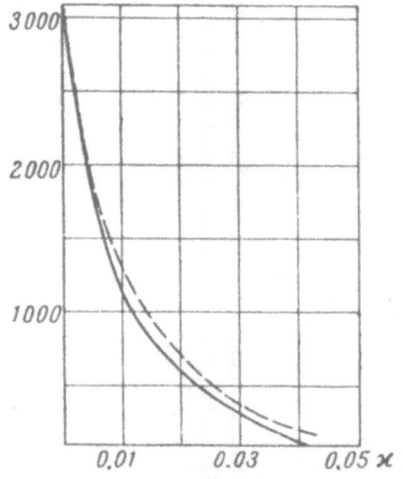

Fig. 43. Calculation of the static sensitivity of tangentially polarized barium
titanate cylinders by the approximation
and accurate formulae, respectively.

possess the same sensitivity. Figure 44a shows a longitudinally
polarized cubic piezoelectric element (1 cm on a side) and a
disc-shaped element of 1 cm thickness of equivalent sensitivity;
in both cases the external force is directed parallel to the direction of polarization. Figure 44b shows the same cube and disc
subjected to all-round lateral pressure; in this case the direction
of deformation is perpendicular to the direction of polarization.
Figure 44c shows a thin-walled, radially polarized cylindrical
shell with rigid end caps; the external pressure acts on the entire
surface of the piezoelectric element. Figure 44d shows a polarized cube of unit side deformed in two directions (perpendicular and parallel to the direction of polarization) and the
equivalent solid radially-polarized cylindrical element of unit
radius subjected to all-round lateral pressure. Figure 44e shows
a cube of unit side subjected to deformation in a direction perpendicular to the direction of polarization, a thin-walled radially polarized spherical shell of unit radius subjected to hydrostatic pressure, a radially-polarized cylindrical shell of unit
radius subjected to high hydrostatic pressure, and the same shell
of unit radius with unshielded ends. Finally, Fig. 44f shows a
cube of unit side, a solid radially polarized sphere of unit radius,
a solid radially polarized cylinder of unit radius, and a disc of
unit thickness all subjected to hydrostatic pressure.

We cannot include tangentially polarized elements in this comparison since in this case the sensitivity
is determined not only by the geometrical dimensions of the shell but also by the number of electrodes used.
Figures 44c and 44e clearly show that thin-walled shells result in incomparably better utilization of the ceramic piezoelectric material than any form of solid ceramic element.

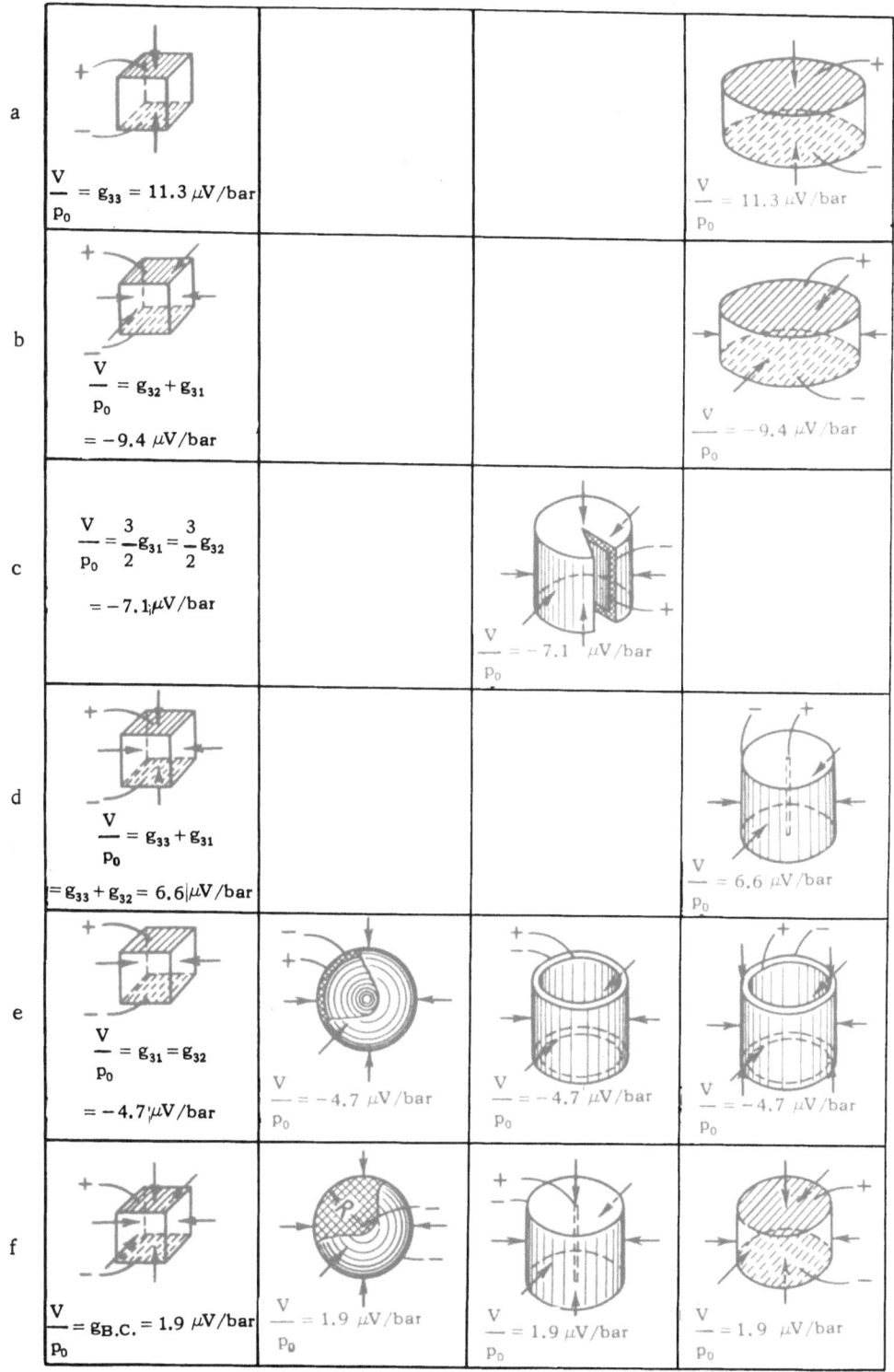

Fig. 44. Comparison of the sensitivity of spherical and cylindrical piezoelectric elements with the sensitivity of the simplest types of element, i.e., a cube or plane slab (radius of cylinder or sphere equal to the side of the cube or the thickness of the slab).

5. Cylindrical and Spherical Sound Receivers in Which the Piezoelectric Element Is Incorporated in the Receiver Shell

As already stated it is possible in principal to construct sound receivers in which the shell (acting as the acoustic antenna) can be made of any material, e.g., metal, glass, plastic, etc., and fulfills the function of a mechanical transformer, the piezoelectric element itself being incorporated in this shell. With sufficiently large external dimensions of the cylinder or sphere and small wall thickness it is possible to obtain large coefficients of mechanical transformation; at the same time, in view of the small dimensions of the incorporated element, the latter exerts no appreciable effect on the vibration of the shell.

On the basis of these considerations we investigated a number of different types of sound receivers in which piezoelectric elements of barium titanate ceramic were incorporated in the shell. The principal data are given in Table 14.

The piezoelectric elements were incorporated in the shells as shown in Fig. 22.

A metallic shell can of course also act as an electrostatic screen. The use of a glass shell (or any other electrically insulating shell) requires an additional screen which in our case consisted of a layer of silver deposited by chemical means on the internal surface of the shell.

Figure 45 shows the directivity characteristics of a sound receiver with a glass spherical shell 5 cm in diameter. The diameter of the built-in piezoceramic element was 1 cm. The reference directions are in-

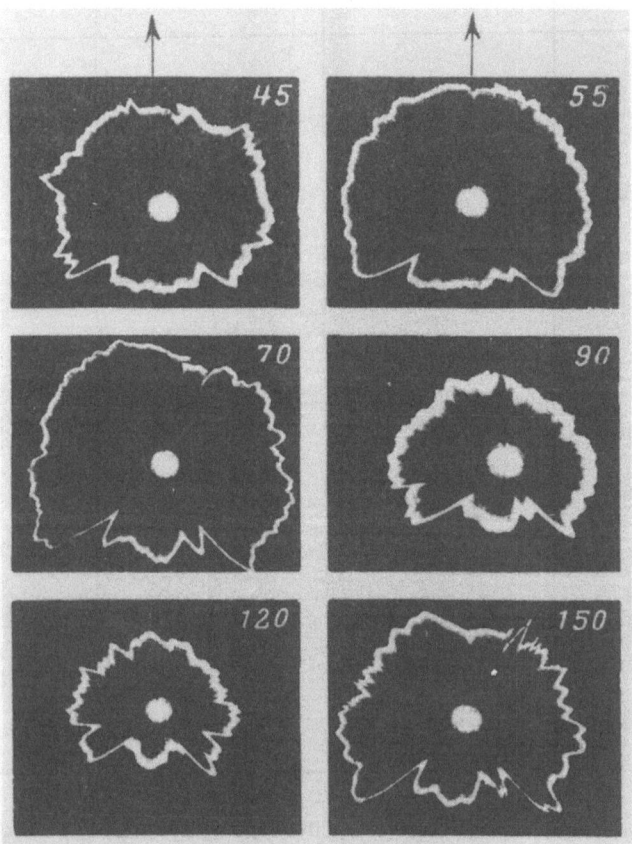

Fig. 45. Directivity characteristics of a sound receiver with a spherical shell incorporating a sensitive piezoceramic element. The figures give the frequencies in kc.

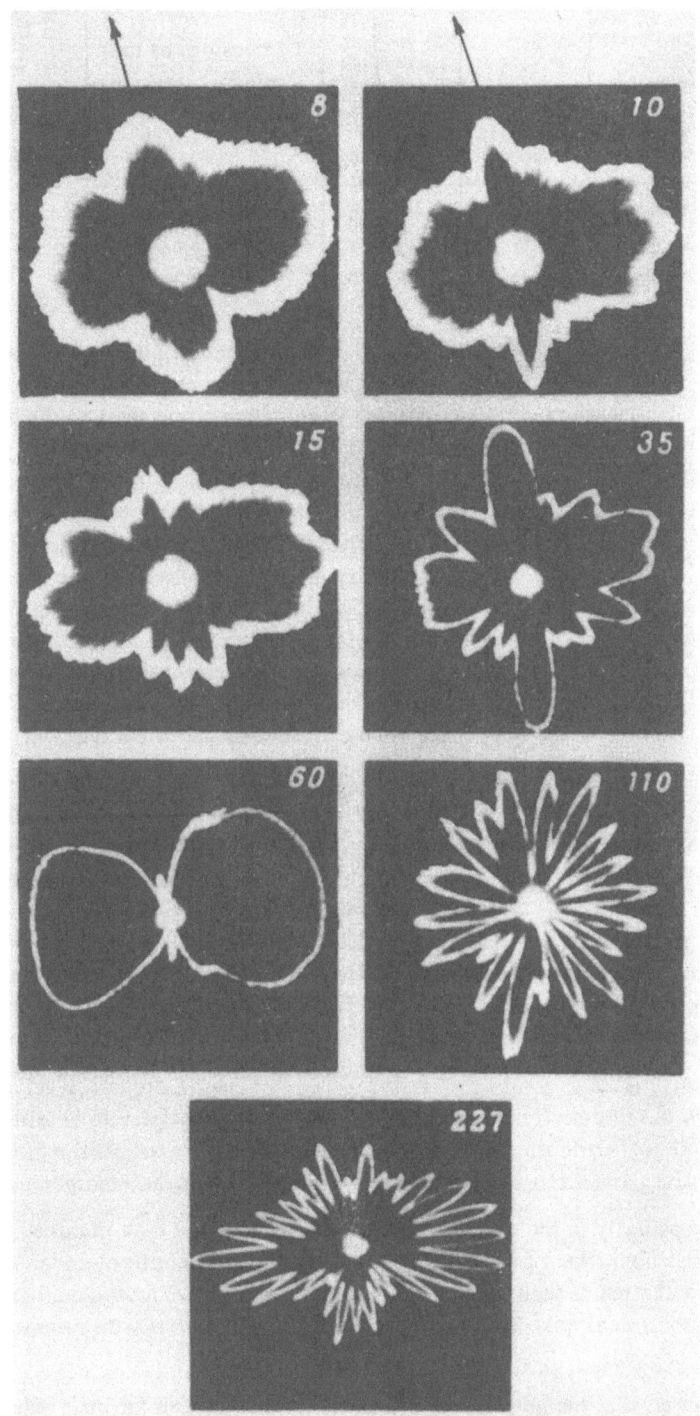

Fig. 46. Directivity characteristics of a sound receiver with a cylindrical shell incorporating a piezoceramic element. The figures give the frequencies in kc.

TABLE 14

Form of the shell	Material	Shell dimensions		Dimensions of the piezoelectric element	Capacity of the piezoelectric element	Transformation coefficient	Theoretical sensitivity of the receiver, μV/bar
		Diameter, mm	Wall thickness, mm				
Cylindrical	Steel	38	2.0	2 × 6 × 40 mm	100	9.5	57
Spherical {	Glass	50	1.5 }	Diameter 10 mm,	37	7.5	37
{	Glass	29	1.5 }	thickness 2 mm	37	4.8	37

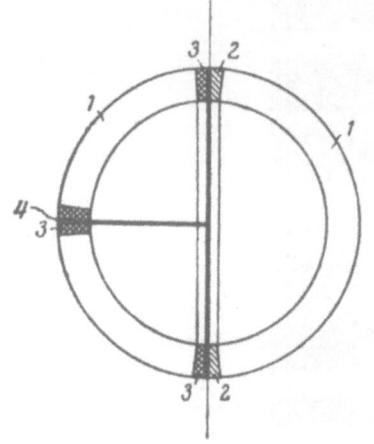

Fig. 47. Example of a receiver consisting of a spherical shell combined with a sensitive ceramic element. 1) Steel hemispheres; 2) equatorial region consisting of a ring of barium titanate ceramic; 3) electrically insulating material; 4) electrical lead.

dicated by the arrows and correspond to the position of the receiver. The piezoelectric element was inserted in the rear surface of the shell at the point of attachment of the shell to the arrangement for rotating the shell during the recording of the directivity characteristics. It is clear that at frequencies from 45 to 150 kcs the characteristic is close to circular in the forward hemisphere and is considerably distorted in the rear hemisphere. This is apparently due to violation of constructional symmetry arising from the presence in the symmetrical receiver shell of a piezoelectric element of different material possessing different elastic constants. In addition, any defects or inhomogeneities arising from imperfect union between the piezoelectric element and the shell will undoubtedly introduce distortion into the directivity characteristics of the sound receiver.

Figure 46 shows the directivity characteristics in water of a sound receiver with a steel cylindrical shell with a diameter of 3.8 cm and a length of 4 cm. The dimensions of the two incorporated elements which extended the full length of the cylinders as shown in Fig. 22b were 4 × 0.6 cm with a radial thickness of 0.17 cm. The reference directions are indicated by the arrows. We see that the complete splitting of the shell by the insertion of two piezoelectric elements together with probable imperfections in the joints between the shell and the elements leads to much greater distortion of the directivity characteristics than in the case of the spherical shell.

In order to improve the directivity characteristics it is clearly necessary to obtain a more rigid bond between the shell and the piezoelectric element and to select materials for the shell and piezoelectric element which possess similar elastic parameters and to reduce the dimensions of the incorporated element.

Despite the fact, especially in the case of spherical shells, that the data obtained showed that the conception of a sound receiver consisting of a shell incorporating a piezoelectric element was theoretically valid, nevertheless practical realization of such sound receivers encountered serious technological difficulties. Therefore, in what follows we shall deal only with nondirectional sound receivers with thin-walled shells made entirely of barium titanate ceramic.

It is possible, however, that techniques will eventually be developed for the production of shells incorporating piezoelectric elements.

In this case, the spherically symmetric construction shown schematically in Fig. 47 would be possible. This consists of two steel hemispheres rigidly connected by an electrically insulating plate with an equatorial region in the form of a ring of barium titanate ceramic. This type of construction would enable reliable and homogeneous bonding and would be suitable for low-frequency nondirectional sound receivers of large diameter.

We investigated a number of different types of spherical piezoceramic radially polarized sound receivers with electrodes deposited over the entire internal and external surfaces of the spherical shell. The requirement for minimum disturbance of the spherical symmetry meant that the aperture for the lead from the internal electrode had to be kept as small as possible. Since the preparation of a complete hollow sphere of uniform wall thickness at any cross section, containing a small aperture for the lead from the internal electrode, would encounter considerable technological difficulties, we prepared two hemispheres (Fig. 48, 1) and assembled these to form a spherical element with a diametral joint 4. The hemispheres were formed on an optical bench with an accuracy of 0.05 to 0.1 mm; the electrodes 2, 3 were deposited on the internal and external surfaces of the hemispheres by repeated silver firing. The hemispheres of the experimental models were bonded by means of a carbinol adhesive or a carbinol adhesive containing a solid filler (talc, barium titanate powder, etc.). Prior to assembly the hemispherical units were simultaneously radially polarized under identical conditions. All the basic measurements required to determine the properties of sound receivers of this type of construction were carried out on prototypes assembled in this way.

However, in the final models an improved method of bonding the two hemispheres was used, namely vitreous welding. The lead from the internal electrode was brought out through a vitreous insulating bush welded into the aperture in the ceramic shell. In order to improve the resistance to moisture penetration the entire spherical element was coated with vitreous enamel. The cross section of such a construction is shown schematically in Fig. 48b.

Our experience has shown that this type of spherical piezoelectric element is best constructed in the following order:

1) the two hemispheres are prepared by grinding on an optical bench and the internal and external surfaces are metallized by silver firing. In the case of the internal surface a narrow region along the diametral joint is left free from metallization;

2) thin silver wire leads (0.1 to 0.2 mm in diameter) are soldered to the internal metallized surface; each lead is secured by glass beads (Fig. 48, 8) at one or two points on the internal electrode 3, the electrical connection being achieved by silver firing;

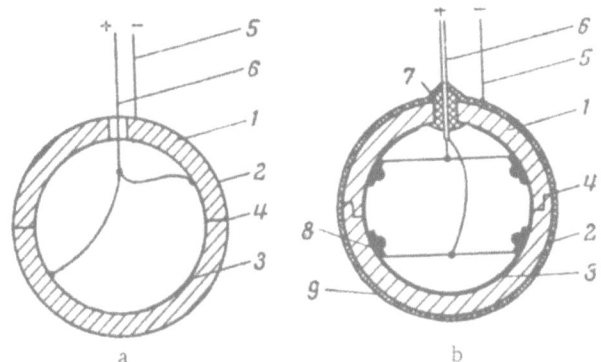

Fig. 48. Ceramic element of a spherical sound receiver. a) For use in air; b) for use in water. 1) Ceramic hemisphere; 2) external electrode; 3) internal electrode; 4) diametral bond; 5) lead from external electrode; 6) lead from internal electrode; 7) vitreous insulating bush; 8) glass bead; 9) external vitreous coating.

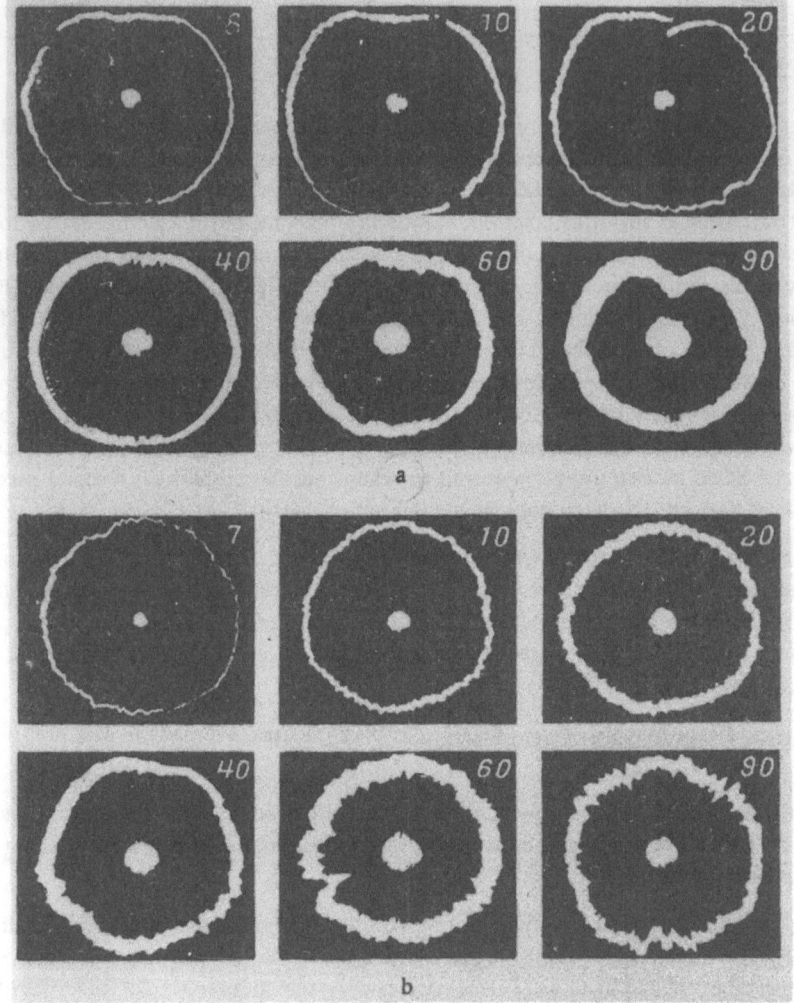

Fig. 49. Directivity characteristics for two spherical sound receivers with
(a) meridional, (b) equatorial joints. The figures give the frequency in kc.

3) the lead to the internal electrode is brought out through an aperture in one of the two hemispheres; the lead takes the form of a silver wire (0.5 to 1.0 mm in diameter, depending on the size of the spherical element). This rigid connection is also used to mount the sound receiver on the connecting element;

4) the two hemispheres are now welded along the diametral joint 4 to form a single spherical piezo-electric element;

5) the entire surface of the sphere (including the joint) is now coated with silver by the silver firing method;

6) the external surface of the sphere is now coated with a waterproof layer of vitreous enamel 9;

7) the assembled piezoelectric element is polarized.

The temperature for reducing the silver from the silver paste should be 100 to 150° below the melting point of the glass used to weld the two hemispheres together and to coat the external surface of the sphere.

Spherical elements assembled from two hemispheres possess two symmetry violating factors, i.e., the output lead from the internal electrode and the welded diametral joint. In the case of underwater sound receivers, the lead from the internal electrode must also be insulated from contact with the water which still further disturbs the constructional symmetry. Two series of experiments were carried out to investigate the

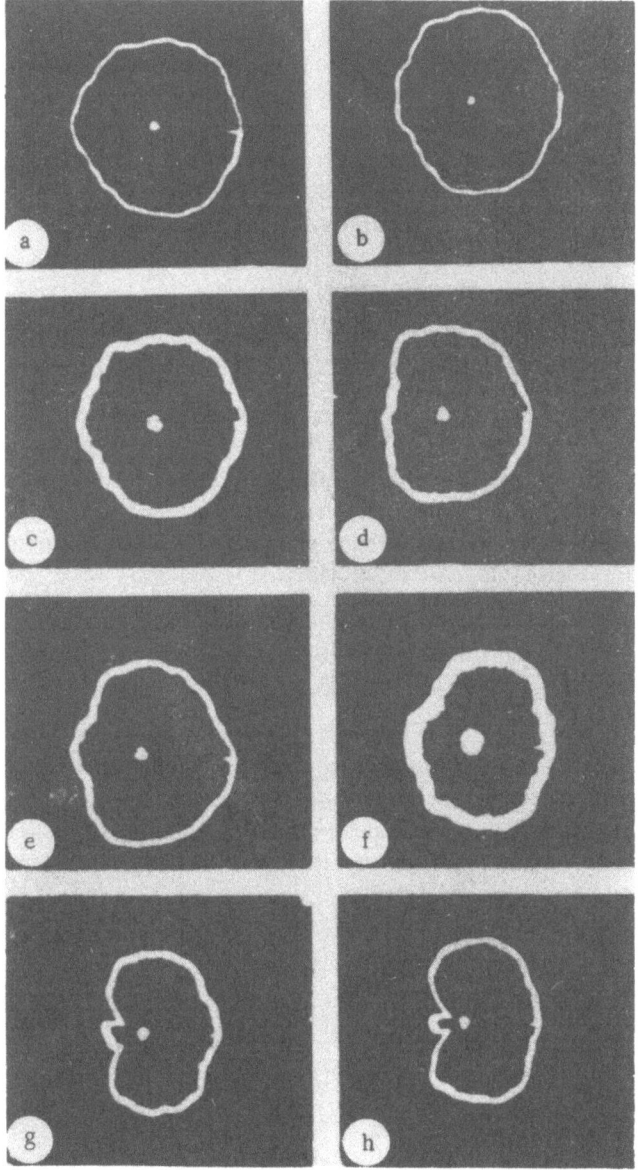

Fig. 50. Directivity characteristics of a spherical sound receiver
of diameter D; part of the receiver surface is covered with a screen
of diameter D'. a) Spherical sound receiver (D ≈ λ); b) brass
(D' ≈ 0.25 λ); c) transparent plastic (D' ≈ 0.35 λ); d) transparent
plastic (D' ≈ 0.8 λ); e) cork (D' ≈ 0.35 λ); f) lead (D' ≈ 0.8 λ);
g) microporous rubber (D' ≈ 0.7 λ); h) cork (D' ≈ 0.8 λ).

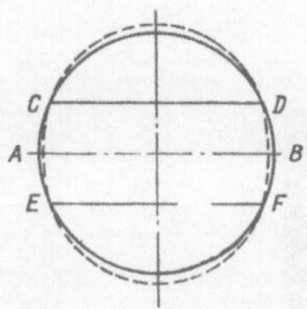

Fig. 51. Vibration mode of
a shell with an imperfect
diametral joint.

effects of the joint and the hermetic sealing of the lead from the internal electrode. We first recorded the directivity characteristics of two spherical sound receivers of the simplest construction in one of which the joint in the ceramic element coincided with the plane which included the lead and which we arbitrarily designate the meridional plane while in the other the joint coincided with the equatorial plane. The directivity characteristics were recorded in the equatorial plane. The directivity characteristics for sound receivers with meridional and equatorial joints are shown in Figs. 49a and 49b, respectively, and are easily seen to be practically identical at the same frequencies. This indicates that the joint has no appreciable effect on the receiving properties of a ceramic shell. In fact, the joint has no effect on the directivity characters irrespective of whether the joint is effected by vitreous welding or by the less efficient method of jointing with the aid of a carbinol adhesive.

The second series of experiments was carried out to investigate the effect of additional elements required to provide a reliable watertight seal for the lead from the internal electrode.

Figure 50a shows the starting directivity characteristic recorded in the equatorial plane at a frequency of 40 kc for a sound receiver 3.4 cm in diameter while Figs. 50b to h give the directivity characteristics at the same frequency with the addition of a disc of nonceramic material attached by adhesive at the equator of the sphere. The thickness of the discs was 1.5 mm. The diameters expressed as fractions of the wavelength in water are given for each case in Fig. 50 which shows that the presence of additional elements at the surface of the sphere exerts a considerable effect on the directivity characteristics.

From the results of these experiments we may conclude that in the case of nondirectional sound receivers we can use spherical elements with equatorial joints; in order to reduce distortion of the directivity character- istics in the plane containing the lead from the internal electrode, the dimensions of the components of the hermetic seal should be as small as possible.

We shall now consider the resonance properties of spherical shells used in the construction of nondirec- tional ceramic sound receivers. A sound receiver is characterized by its working frequency band, i.e., the fre- quency band within which the sensitivity of the receiver remains constant within given limits. This frequency band possesses an upper limit coinciding with the first resonance frequency. The method used to join the hemi-

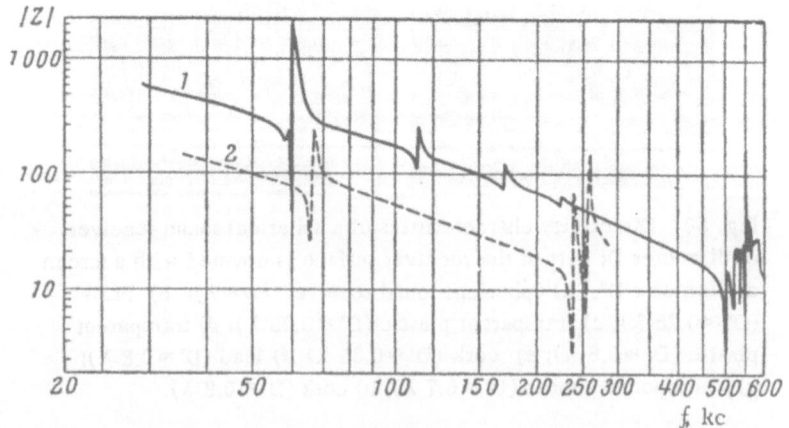

Fig. 52. Modulus of the electrical impedance of two spherical sound re- ceivers with an adhesive-type diametral joint. External diameter D = 50 mm; 1) wall thickness 5 mm; 2) wall thickness 10 mm.

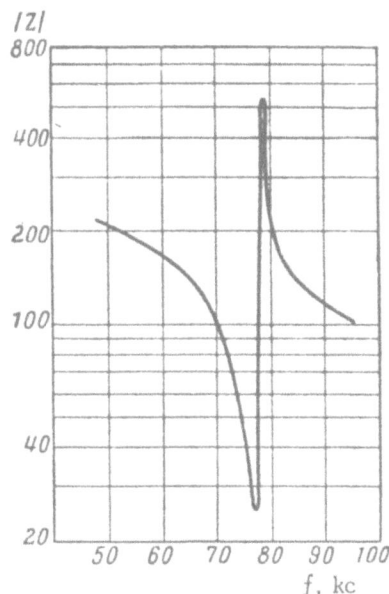

Fig. 53. Modulus of the electrical impedance of a spherical sound receiver with a vitreous welded diametral joint. External diameter D = 30 mm.

spheres together has a considerable effect on the value of this frequency. In order to investigate this effect we determined the vibration amplitude distribution over the surface of the shell with the aid of a special piezoelectric probe. The experiments were carried out in air with different ultrasonic voltages fed to the electrodes of the sound receiver. In the case of shells joined together by vitreous welding, the first resonance in the frequency characteristic corresponded to the zero mode of vibration of the shell. With a weaker form of joint (e.g., as obtained by the use of a carbinol adhesive) the first resonance corresponded to the vibration mode shown schematically in Fig. 51 in which two identical hemispheres are joined together by a diametral joint AB. In this case the vibrations possess axial symmetry characterized by the presence of two nodal lines (CD and EF, diameters of the nodal circles). In this resonance case the corresponding zero mode does not appear at all and has no effect on the frequency characteristics of the sound receivers. Since for the same shell geometry the modal frequency corresponding to Fig. 51 is higher than the zero mode frequency, we might assume that the weaker adhesive joint is more favorable since it permits displacement of the first resonance to a higher frequency, thereby extending the working frequency band. However, this applies only to sound receivers destined for use in air for the measurement of small static pressures; in most cases it is necessary to use a strong joint even at the expense of some reduction of the upper limit of the working frequency band.

As stated earlier, the limit frequency can be conveniently determined from measurements of the electric impedance of the sound receiver; in fact up to the first resonance the sensitivity of the receiver differs only slightly from the static sensitivity, and if we know the static sensitivity and the location of the first resonance, we can determine the working frequency band of the receiver. In order to take account of the effect of diffraction phenomena on the frequency characteristic the latter must be recorded under the actual conditions of operation, i.e., in air or in water.

Figure 52 gives examples of the frequency dependence of the electrical impedance of spherical sound receivers for which the joining was achieved with a carbinol adhesive. For curve 1 the receiver was 50 mm in diameter and possessed a wall thickness of 5 mm; curve 2 relates to a receiver of the same diameter with a wall thickness of 10 mm.

Figure 53 shows the corresponding curve for a sound receiver in which the hemispheres are joined together by vitreous welding. In this case the diameter of the sound receiver was 30 mm; the wall thickness was

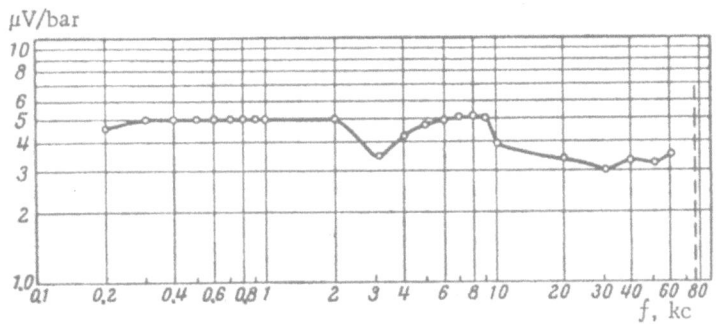

Fig. 54. Frequency characteristic of a spherical sound receiver. External diameter D = 30 mm.

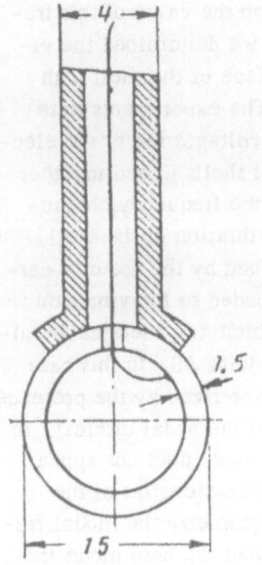

Fig. 55. Sound receiver
with the spherical receiver
element rigidly secured to
the support.

1.5 mm. Finally, Fig. 54 shows a typical frequency characteristic for a spherical sound receiver with a diameter of 30 mm and a wall thickness of 1.5 mm in an aqueous medium. The reduction in sensitivity prior to resonance (the location of the resonance is indicated by a dotted line) is caused by diffraction effects; at lower frequencies the frequency characteristic is flat and the sensitivity is practically equal to the static sensitivity.

We shall now examine a number of design problems of particular importance in the development of nondirectional wide-band sound receivers. Firstly, we must pay attention to methods of attaching the sensitive element to the holder or other intermediate supporting element facilitating connection between the sound receiver and the cable or the preamplifier. In the case of a sound receiver designed for use in air, such a connecting element must provide adequate strength, reliable connection, and no disturbance of the electrical screen. In the case of underwater application there is the added requirement of watertight sealing. Generally speaking, fulfillment of these requirements leads to an increase in the size of the connecting element which may result in distortion of the directivity characteristic in the meridional plane. Faulty design of the connecting element may also lead to distortion of the frequency characteristic.

For example, let us consider the case of a sound receiver in which the spherical element is rigidly secured to a metal tube (Fig. 55). The end of the tube carries a bowl-shaped seating to which the spherical element is secured by means of a carbinol adhesive. Figure 56a shows the directivity characteristics in the equatorial plane for such a sound receiver with a piezoelectric element 1.5 cm in diameter and a wall thickness of 1.5 mm. The diameter of the metal tube was 4 mm, the length of the tube being 35 cm. Figure 56b shows the directivity characteristics in the meridional plane for the same frequencies. It is clearly seen that these characteristics are highly distorted and that distortion is also evident in the directivity characteristics in the equatorial plane. At almost all frequencies, the directivity characteristics in the meridional plane possess sharply expressed maxima at angles of incidence of the sound wave of ±65°30' relative to the axis of the support. Detailed investigation showed that the presence of these maxima and the generally high degree of distortion of the directivity characteristics in the meridional plane were due to the fact that the tubular support acted as an additional sound antenna, thus disturbing the directional properties of the spherical element. Rigid connection of the sound receiver to the support also affects the frequency characteristic of the sound receiver. For example, Fig. 57 shows the frequency characteristics of the electrical impedance for the piezoelectric element alone (dotted curve) and mounted on the support (solid line curve). As would be expected, the presence of the support lowers the resonance frequencies and reduces the quality in resonance.

The above-mentioned examples of distortion of the directivity characteristics and possible additional disturbances in the frequency characteristic of the sound receiver indicate the undesirable effects associated with a rigid connection between the piezoelectric element and the support. Therefore, we dispensed with rigid mountings and used a type of mounting in which the piezoelectric element is isolated from the support by means of a flexible element which acted as a mechanical decoupling filter. For example, Fig. 58 shows the method used to connect a large diameter piezoelectric element directly to the output cable. The spherical piezoelectric element 1 is secured to the cable by means of a sufficiently flexible rubber bush of conical form 6 which is secured by rubber adhesive to the dielectric cable 7. The base of the rubber bush 6 is secured by adhesive to the external metallized surface of the piezoelectric element. An insulating bush 2 of textolite (or some other insulating material) extends into the rubber bush thus forming a watertight seal for the lead from the internal electrode.

The directivity characteristics of a spherical sound receiver attached to the cable at the meridional plane as described above are practically identical with those shown in Fig. 45 and may be considered satisfactory although the distortion introduced by the mounting in the rear hemisphere is still considerable.

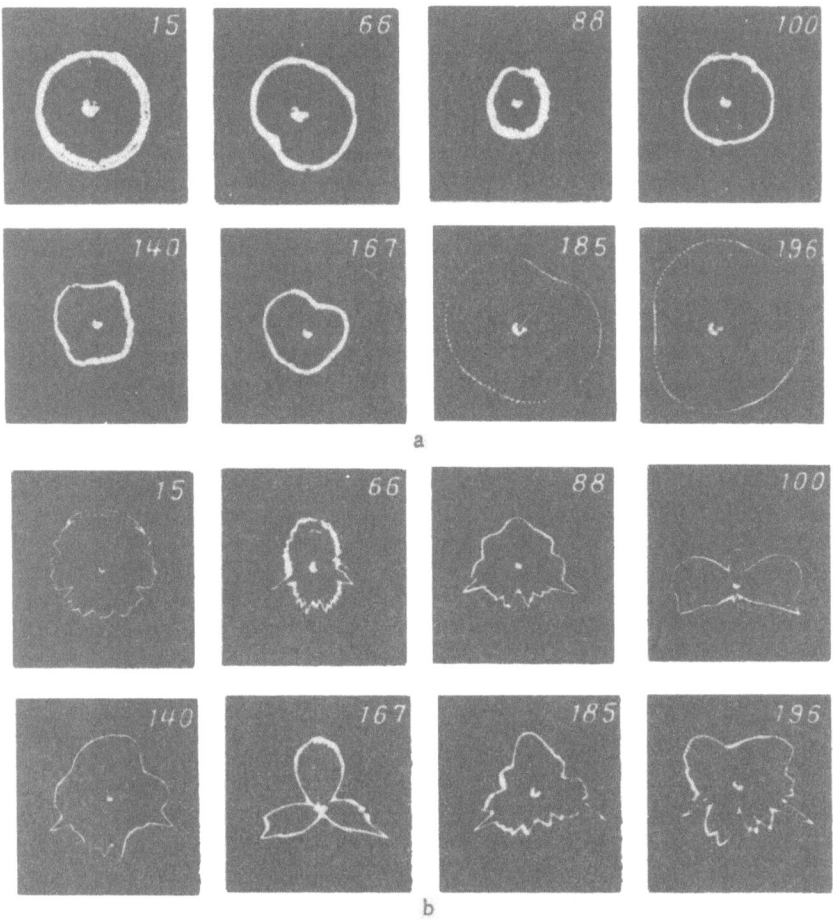

Fig. 56. Directiviiy characteristics in (a) equatorial (b) meridional planes for a sound receiver in which the spherical element is rigidly secured to the support. External diameter of the piezoelectric element 1.5 mm. The figures give the frequencies in kc.

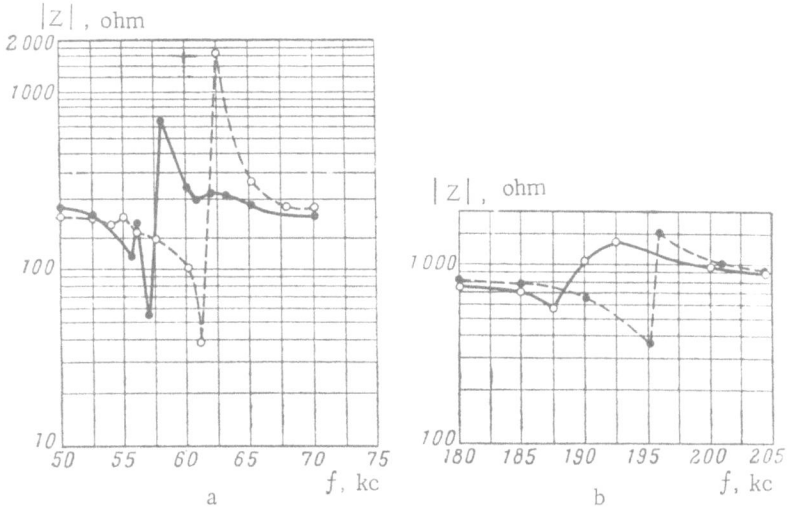

Fig. 57. Frequency characteristic of the modulus of electrical impedance of spherical piezoelectric elements and sound receivers. a) External diameter of the piezoelectric element 50 mm; b) external diameter 15 mm.

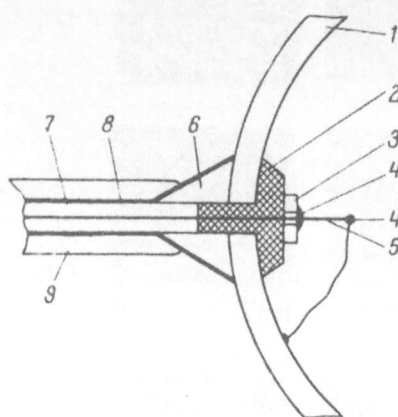

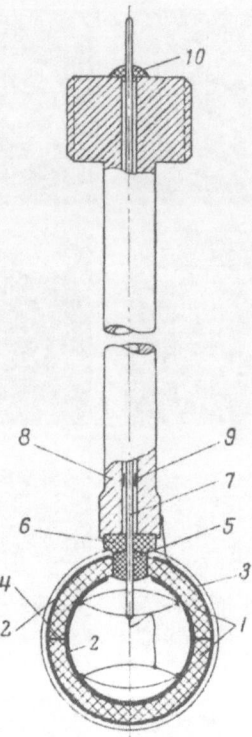

Fig. 58. Method of securing the spherical element of the sound receiver to the cable. 1) Spherical element; 2) textolite bush; 3) metal disc; 4) tin solder bead; 5) central conductor of the coaxial cable; 6) conical rubber bush; 7) dielectric of the coaxial cable; 8) cable screen; 9) external covering of rubber or plastic.

Fig. 59. Method of securing a small spherical element to the support. 1) Ceramic piezoelectric element; 2) metallic coating; 3) external coating of vitreous enamel; 4) vitreous welded joint; 5) vitreous bush; 6) rubber decoupling element (forming watertight joint); 7) rod-type lead from the internal electrode; 8) tubular support; 9) insulating bushes; 10) lead secured with carbinol adhesive.

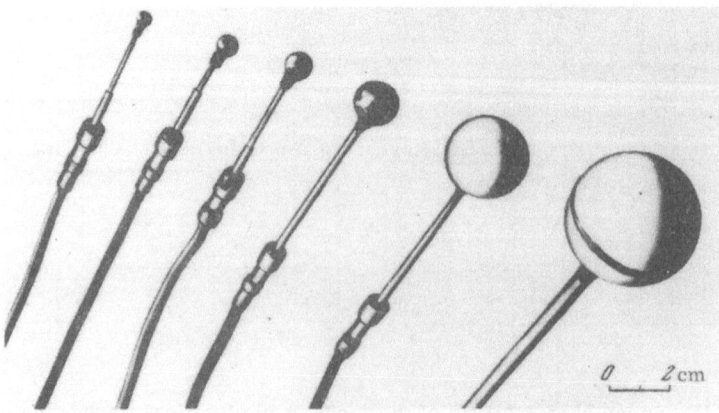

Fig. 60. Set of spherical measuring sound receivers.

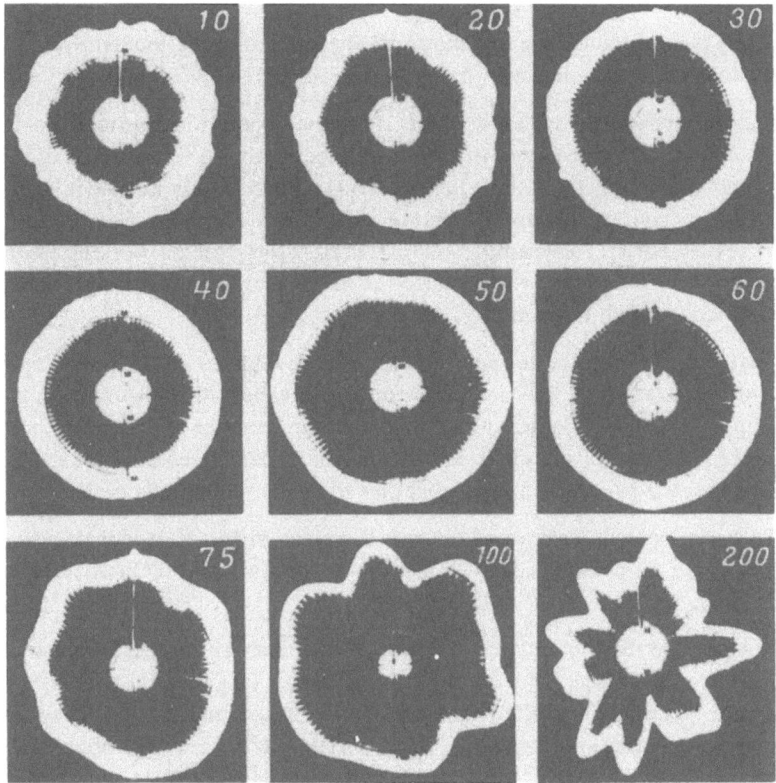

Fig. 61. Directivity characteristics in the equatorial plane for a spherical ceramic sound receiver. External diameter D = 50 mm. The figures give the frequencies in kc.

In the case of small spherical elements used in high-frequency sound receivers, it is not desirable to attach the element to the end of the coaxial cable since in this case the presence of the cable causes marked distortion of the acoustic field. Therefore, it is necessary to use different types of support and decoupling element.

A method of supporting a small spherical element is shown in Fig. 59.

In this case, the spherical element 1 is assembled by means of a vitreous welded diametral joint and the lead from the internal electrode takes the form of a silver rod which passes through the vitreous insulating bush 5. The procedure for constructing this type of piezoelectric element was described previously. The assembly of sound receivers with the type of piezoelectric element shown in Fig. 61 requires special accuracy and skill on the part of the technician. A particularly high degree of skill is required in the assembly of sound receivers with spherical elements 4 to 5 mm in diameter or smaller.

We constructed a series of wide-band, nondirectional sound receivers of this type with spherical element diameters of 4 to 50 mm (Fig. 60). The measured static sensitivities corresponding to external diameters D = 15, 30, and 50 mm were 1.34, 2.9, and 7.9 μV/bar, respectively. There was very little difference between the measured and theoretical static sensitivities.

Some idea of the directivity characteristics of these sound receivers can be obtained from Fig. 61 which shows the equatorial directivity characteristics for a sound receiver with an external diameter of 50 mm.

We note that with completely satisfactory directivity characteristics in the equatorial plane the directivity characteristics in the meridional plane can only be regarded as nondirectional in the forward hemisphere since considerable distortion is still introduced into the rear hemisphere due to the presence of the support.

One advantage of the type of sound receiver shown in Fig. 60 is that the receiver remains intact in the event of damage to the cable (breakage, leakage, etc.).

However, in the case of underwater sound receivers, increased operational reliability is obtained if the piezoceramic element is completely vulcanized in rubber together with the termination of the cable. Our experiments showed that a rubber coating 1 to 2 mm in thickness had practically no effect on the sensitivity of wide-band, spherical, radially polarized ceramic sound receivers with diameters from 10 to 50 mm within the limits of the working frequency band. For example, the mean sensitivity in the working frequency band for a sound receiver in which the spherical element was totally enclosed in a rubber coating was 5 μV/bar and 8 μV/bar for diameters of 30 and 50 mm, respectively.

Finally we note that in addition to radial polarization it is possible to use tangential polarization. In order to obtain tangential polarization we can use the electrode arrangement shown in Fig. 35 in which the metallized layer shows up as a bright background against the darker circular regions of the nonmetallized areas of the ceramic. This layer forms one of the electrodes. The small metallized areas at the center of each darker circle which are connected by soldered leads together form the second electrode.

Since the electrodes are located on the same (internal) surface of the shell it is necessary to select a polarizing field (600 to 700 V/cm for polarization with heating to the Curie point) in accordance with the distance between the electrodes.

The electrodes can be deposited not only on the internal or external surfaces but also on both surfaces of the spherical shell. In the case of very thin shells (0.2 to 0.3 mm) it is preferable to have one system of electrodes (e.g., the central electrode) deposited on the internal surface and the second system of electrodes on the external surface of the spherical shell. In this case, only two leads are required outside the sphere. With very thin-walled shells the sensitivity of such tangentially polarized piezoelectric elements may reach hundreds of μV/bar for shell diameters of 3 to 5 cm. Tangentially polarized spherical piezoelectric elements will undoubtedly find application in the design of sound receiver-microphones for use in air.

7. Cylindrical Wide-Band Sound Receivers with Thin-Walled Shells of Barium Titanate Ceramic

A number of acoustic problems can be solved with the aid of a sound receiver with a circular directivity characteristic in one plane. Therefore, a number of sound receivers were constructed with cylindrical elements of barium titanate ceramic.

It was shown previously that piezoelectric elements in the form of ceramic cylinders permit the use of three methods of polarization and three different sets of conditions at the end surfaces of the cylinder with correspondingly different types of cylindrical sound receiver.

We developed two types: 1) receivers with elements in the form of radially polarized ceramic cylinders with end caps and 2) receivers with tangentially polarized cylinders with the end faces of the cylinder screened from the effect of the sound pressure.

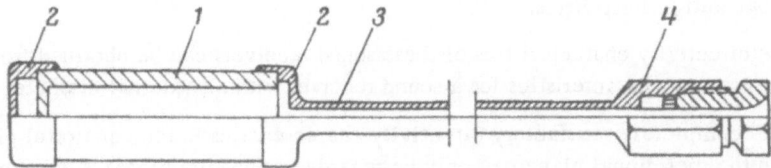

Fig. 62. Cylindrical sound receiver with a rigidly mounted, radially polarized element. 1) Cylindrical piezoelectric element; 2) metal seating; 3) tubular support; 4) connector.

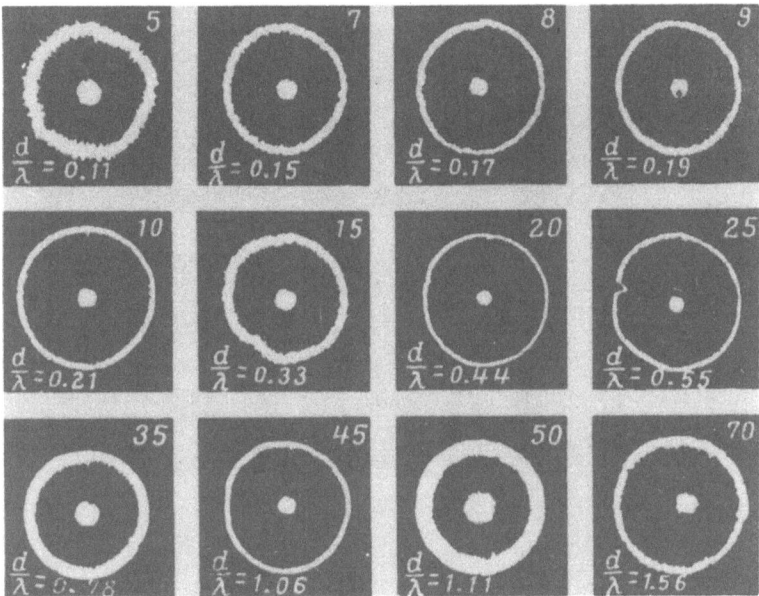

Fig. 63. Directivity characteristics of a cylindrical radially polarized sound receiver of barium titanate ceramic. The characteristics were recorded in the plane perpendicular to the axis of the cylinder. The figures in the top right-hand corner give the frequency in kc.

In the first case the piezoelectric elements possessed high capacity which enabled the design of sound receivers of very small dimensions with a flat frequency characteristic over a wide frequency band.

In the second case the capacity of the piezoelectric element was smaller, but on the other hand the sensitivity could be very high.

As in the case of spherical sound receivers, the method of securing the cylindrical element to the support affected the properties of the sound receiver. Figure 62 shows the construction of a rigidly mounted, cylindrical sound receiver with a radially polarized piezoelectric element.

The cylindrical piezoelectric element 1, metallized internally and externally to permit radial polarization, is secured in the metal seatings 2 by hot shrinking. These seatings act as end cap receivers of the acoustic field. One of these caps forms the end of the tubular support 3 through which is passed the lead from the internal electrode. The lead is brought out through a watertight connecting sleeve 4 with a rubber packing gland. This type of sound receiver can only be used with an unbalanced amplifier since the external metallization of the piezoelectric element also acts as a screen which is electrically connected to the metal components 2, 3, and 4. As stated, the seatings are shrunk on to the ends of the piezoelectric element and further secured by soldering with Wood's metal thus providing effective watertight sealing of the complete piezoelectric element.

The directivity characteristic of this type of receiver in the plane perpendicular to the axis is almost perfectly circular over a wide range of frequencies. This is evident from Fig. 63 which shows the directivity characteristic for a cylindrical sound receiver of the given type with an external diameter of 3.2 cm, an axial length of 8.1 cm, and a wall thickness of 5 mm recorded in an aqueous medium. In some cases, the directivity characteristics in the plane including the axis (Fig. 64) are completely satisfactory but in other cases secondary maxima are clearly evident. These secondary maxima are to some extent due simply to the natural characteristic of a cylindrical piezoelectric element of finite length, but the most important secondary maxima are undoubtedly caused by the effect of the mounting. However, this disadvantage is not so important in the case of cylindrical sound receivers as in that of spherical receivers.

We employed the experimental sound receiver shown in Fig. 65 to investigate the effect of the mounting on the frequency characteristics of a cylindrical sound receiver, in which the piezoelectric element was

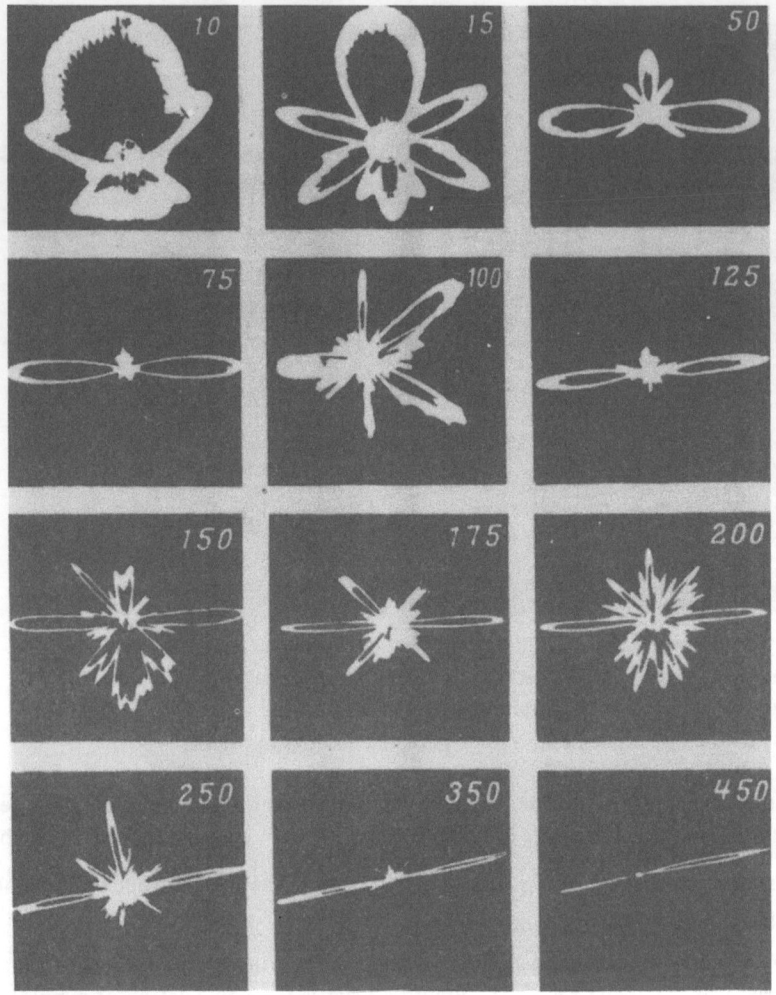

Fig. 64. Directivity characteristics of a cylindrical sound receiver with a
rigidly mounted, radially polarized ceramic element. The characteristics
were recorded in the axial plane of the cylinder. The figures give the
frequency in kc.

mounted as "softly" as possible. For this purpose the piezoelectric element was mounted in two rubber supports 2 and 3 one of which simply covered the end face of the cylinder while the other acted as a flexible decoupling element in the attachment of the sound receiver to the cable.

The directivity characteristics in the plane perpendicular to the axis of cylindrical sound receivers with either rigid or flexible connection of the piezoelectric element differ only slightly from circular right up to very high frequencies (350 to 450 kc) when the dimensions of the sound receiver are considerably larger than the wavelength. For example, with a diameter $D \approx 6\lambda$ and a frequency of 300 kc, the directivity characteristic departs by not more than ±15% from the circular, i.e., a discrepancy of ~1.2 db. In the same way, sound receivers with rigid and flexible mountings differ only slightly in the angular width of the principal lobe. The width of the principal lobe is in good agreement with the theoretical value as indicated by Fig. 66 which shows the theoretical dependence of the width of the principal lobe on frequency (dotted curve) for a cylindrical of finite length; the points indicate the corresponding experimental values obtained with a rigidly mounted piezoelectric element.

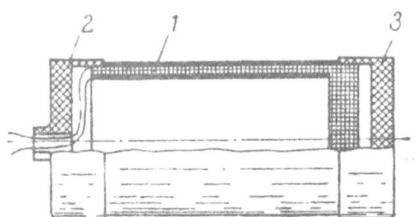

Fig. 65. Cylindrical sound receiver with flexible attachment of the piezoelectric element to the cable. 1) Ceramic, 2 and 3) rubber.

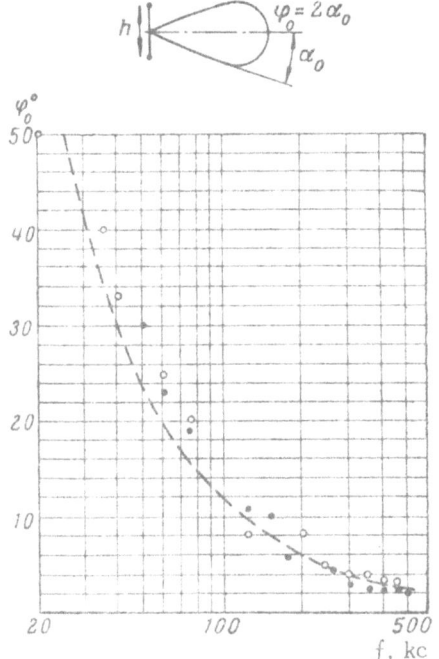

Fig. 66. Width of the principal lobe of the directivity characteristic of a cylindrical sound receiver in the axial plane.

The experimental sound receiver shown in Fig. 65 was used to determine the effect of the type of mounting on the frequency characteristic of the sound receiver. Figure 67a shows the frequency dependence of the sensitivity (in the plane perpendicular to the axis in an aqueous medium) for three identical sound receivers with rigidly mounted piezoelectric elements. Figure 67b shows the corresponding characteristic for an identical piezoelectric element attached to the cable by means of a flexible connection.

It is easily seen that with a flexible mounting the frequency characteristic is much flatter right up to the onset of resonance than in the case of a rigid mounting. Figure 67b shows two clearly expressed resonances.

A cylindrical shell can possess three types of natural resonance: 1) determined by the mean diameter D_1 of the shell (the condition for a first principal resonance of this type is $\pi D_1 = \lambda$); 2) determined by vibrations of the cylinder in the direction of the height h (the condition for a first principal resonance is $h = \lambda/2$); 3) related to vibrations in the direction of the wall thickness δ (the condition for a first principal resonance is $\delta = \lambda/2$). However, all three types of natural vibration are not produced to the same extent by the different methods of polarizing the cylindrical shell. In particular, in the case of radial polarization all three types of resonance are produced but the extent of the working frequency band is limited by either the longitudinal or the radial resonance. In the special case shown in Fig. 67b which indicates the theoretical locations of the longitudinal f_h and radial f_r resonances, the working frequency band is restricted by the first mentioned resonance.

Figure 67a clearly shows a monotonic increase in sensitivity up to 10 kc, a dip at 15 kc, displacement of the first low-frequency resonance in the direction of lower frequencies (from 25 to 20 kc), and the creation of additional resonance peaks due to the rigid mounting of the piezoelectric element. However, rigidly mounted sound receivers are very reliable and convenient in operation since they possess a high safety factor in regard to mechanical strength and permit measurement at comparatively high hydrostatic pressures (of the order of 5 to 7 atm).

Table 15 gives the parameters of two identical cylindrical receivers with rigidly mounted piezoelectric elements.

The reproducibility of the parameters is good. The sensitivity of these receivers is not high because of the relatively large wall thickness.

In subsequent development of wide-band, radially polarized cylindrical sound receivers we considerably reduced the wall thickness to diameter ratio and used only the type fitted with end caps.

Figure 68 shows a typical example of this type of construction consisting of a cylindrical radially polarized piezoceramic element 1 ends of which are covered with rigid caps 2 (metal or ceramic) which act as sound pressure receivers. One of the caps contains an aperture through which the cable is introduced. This aperture is hermetically sealed by means of the gland 6 located inside the cylindrical space of the sound receiver.

TABLE 15

Dimensions of the piezoelectric element			Capacity, μμf	Resonance frequencies, kc		
Height, mm	External diameter, mm	Wall thickness, mm		Longitudinal resonance	Radial resonance	Thickness resonance
81.3	32	5	8,800	14.45	55.5	435
81.3	32	5	8 200	15.3	55.5	435

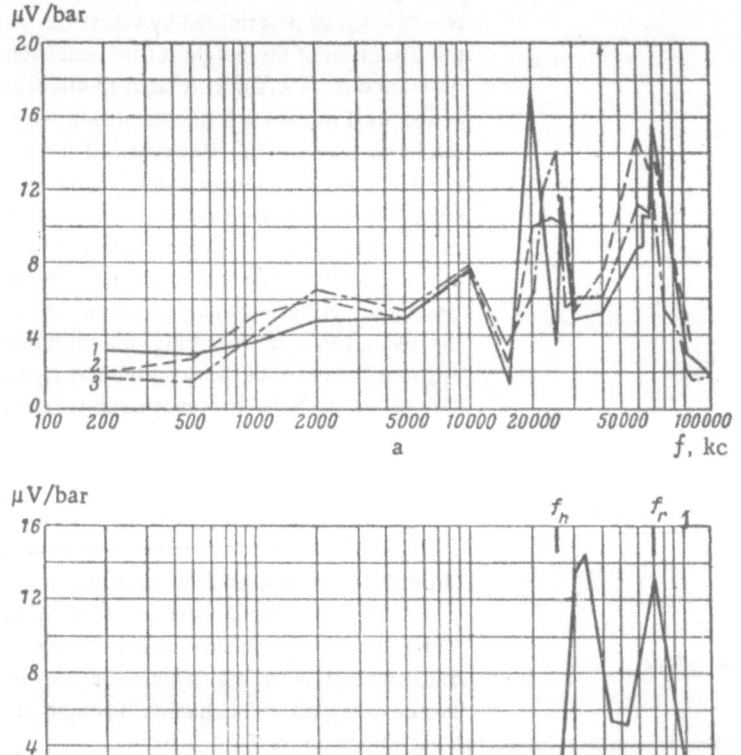

Fig. 67. Frequency characteristics of cylindrical sound receivers of radially polarized barium titanate ceramic. a) Rigidly mounted piezoelectric element; b) flexible connection of the piezoelectric element to the cable. f_n and f_r) longitudinal and radial resonance frequencies.

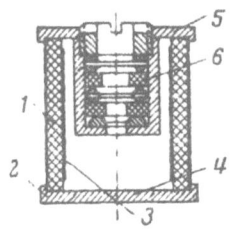

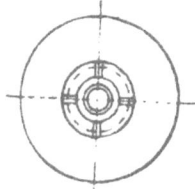

Fig. 68. Construction of a cylindrical ceramic sound receiver fitted with end caps. 1) Piezoelectric element; 2) external metallic coating; 3) internal metallic coating; 4) front end cap; 5) rear metallic end cap; 6) gland.

Figure 69 shows a series of sound receivers of this type, details of which are given in Table 16.

Frequency characteristics recorded in an aqueous medium for cylindrical sound receivers of the given construction with external diameters of 10, 15, 30, and 50 mm are shown in Fig. 70.

In the case of sound receivers in which the diameter of the piezoelectric element is less than 6 mm it is inconvenient to mount the sound receiver directly on the cable and a tubular mount was employed practically identical to that used in the case of flexibly mounted spherical receivers. This type of construction is shown schematically in Fig. 71.

We produced sound receivers of this type with piezoelectric element diameters of 3 to 6 mm an example of which is shown in Fig. 72.

In a number of special cases we continued to mount the sound receiver directly on the cable. Thus, in 1959 we constructed a sound receiver for the Therapeutical Institute of the Academy of Medical Sciences which was designed for introduction into the cavities of the heart for the observation of cardiac noise. The external diameter of this sound receiver was 3 mm. The external appearance of the first prototype of this sound receiver used in the diagnosis of cardiac disease is shown in Fig. 73.

Wide-band cylindrical sound receivers with radially polarized ceramic elements are widely used at the present time together with spherical non-directional sound receivers. As already shown they possess sufficiently high self-capacities and satisfactory sensitivity over a wide range of frequencies together with nondirectional properties in one plane. In many cases, the preparation of cylindrical piezoelectric elements (especially of miniature and subminiature types) is considerably simplified in comparison to the preparation of spherical sound receivers.

We shall now discuss cylindrical sound receivers with tangentially polarized piezoelectric elements. The sensitivity of sound receivers with radially polarized cylindrical piezoelectric elements can only be increased by increase in the external diameter of the element. However, the sensitivity of ceramic sound receivers can be considerably increased by means of tangential polarization.*

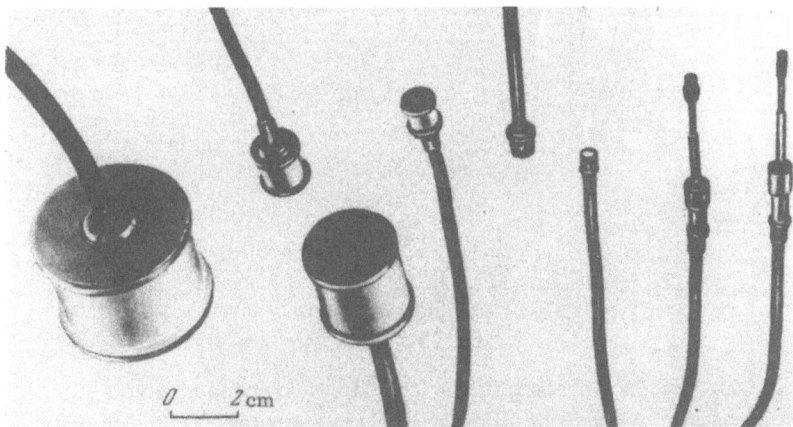

Fig. 69. A set of wide-band cylindrical barium titanate ceramic sound receivers.

---

* In regard to sensitivity and capacity longitudinally polarized cylindrical elements occupy a position intermediate between radially and tangentially polarized elements; therefore they offer no advantage in regard to cylindrical receivers under conditions where the sound pressure is incident on the curved surface of the cylinder.

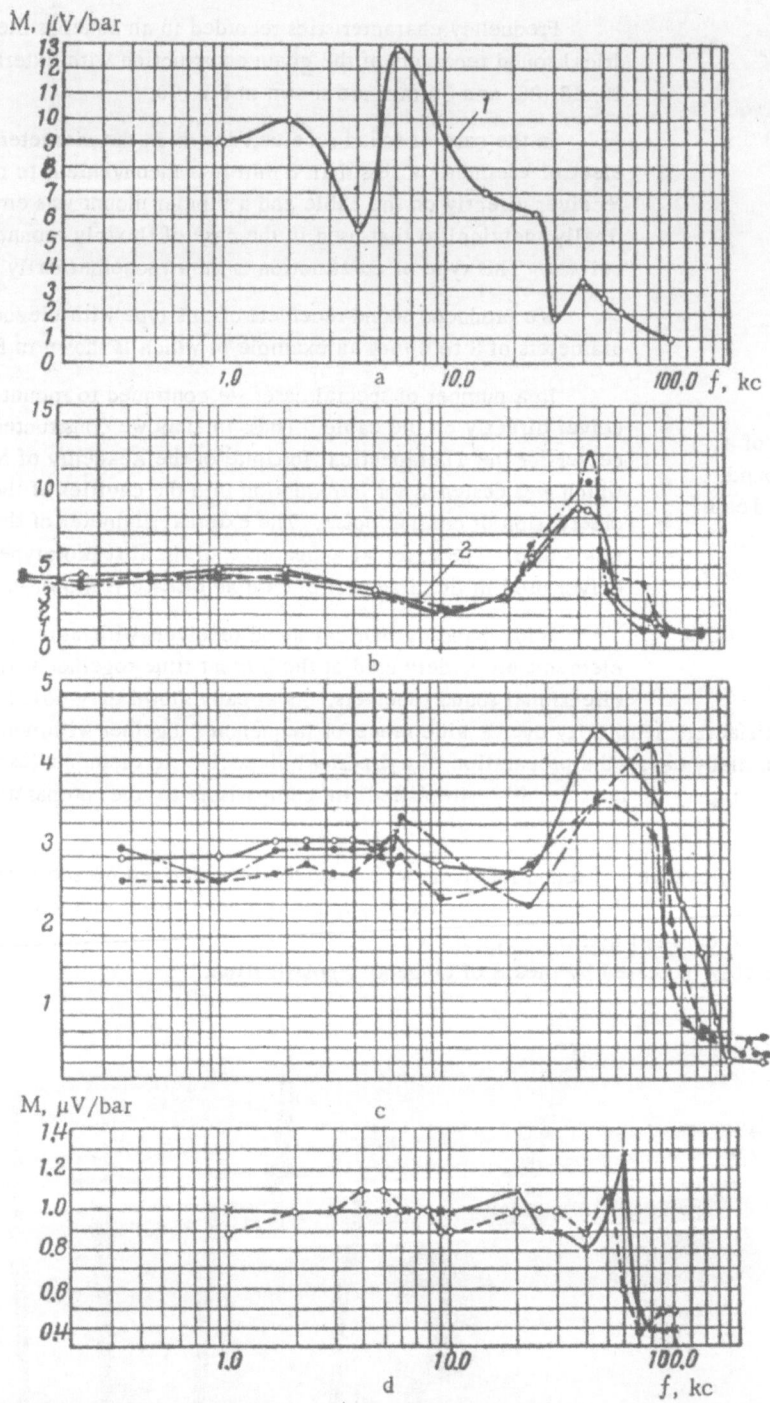

.Fig. 70. Frequency characteristics for cylindrical sound receivers with piezoceramic cylinders of different external diameter. a) 50 mm; b) 30 mm; c) 15 mm; d) 10 mm.

TABLE 16

External diameter of the cylinder, mm	Cylinder height, mm	Wall thickness, mm	Capacity, $\mu\mu f$	Static sensitivity, $\mu V/bar$	First resonance frequency, kc
52	40	3.0	15,000	8.4	20
30	30	1.5	23,000	7.0	50
15	12	1.0	4,500	3.0	100
10	10	1.0	3,000	2.0	150
8	8	0.5	3,500		
6	6	0.5	1,000	0.8	200

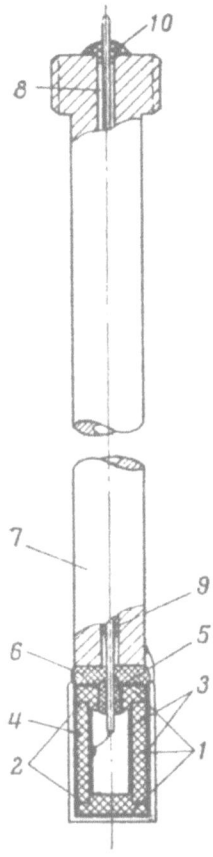

Fig. 71. Method of mounting miniature cylindrical piezo-electric elements on a tubular support. 1) Piezoelectric element of barium titanate ceramic with end caps; 2) vitreous welding; 3) external and internal metallization; 4) external coating of vitreous enamel; 5) vitreous insulating bush; 6) rubber seal; 7) tubular support; 8) rod-type lead; 9) insulators; 10) carbinol adhesive seal.

The construction of a tangentially polarized cylindrical piezo-electric element by subdivision of the cylinder into a number of sections, metallization of the longitudinal end faces of these sections, and the sticking together of the sections to reconstitute the original cylinder is unsuitable for sound receivers with thin-walled shells since the large number of joints destroys the continuity of the shell and undoubtedly adversely affects the directivity characteristic. The above-mentioned method of construction can be used however for cylindrical elements with thick-walled shells containing a large volume of piezoelectric material, e.g., in the case of sound radiators.

Consequently, we employed whole cylindrical shells with the electrodes deposited on either the external or internal surfaces or simultaneously on both surfaces. An example of a cylindrical piezoelectric element with deposited electrodes is shown in Fig. 74a. The electrodes are connected in parallel to form a single electrode. This can be achieved by means of connecting wires as shown in Fig. 74b, but in order to reduce the number of soldered joints it is much better to use the method shown in Fig. 74c where the connecting leads are deposited by silver firing.

The increase in sensitivity due to tangential polarization arises from the following reasons: 1) the use of a thin-walled cylinder gives a high coefficient of mechanical transformation K; 2) use is made of the maximum piezoelectric constant d_{33}; 3) it is possible to increase the distance between the electrodes without changing the wall thickness or diameter of the cylindrical shell.

A certain amount of preliminary discussion is necessary concerning the constructional features of cylindrical sound receivers with tangentially polarized piezoelectric elements. This type of sound receiver cannot possess a continuous external electrode capable of acting as an electric screen. In the case of sound receivers designed for use in air a screen can easily be provided in the form of a cylindrical grid transparent to sound located at a sufficient distance from the electrodes deposited on the surface of the element so that in this case any simple type of construction can be used with electrodes deposited on the internal or external surfaces of the element or on both surfaces simultaneously.

The conditions are different in the case of underwater sound receivers. The aqueous medium unavoidably forms an additional electrode which, if the electrodes are deposited on the internal surface of

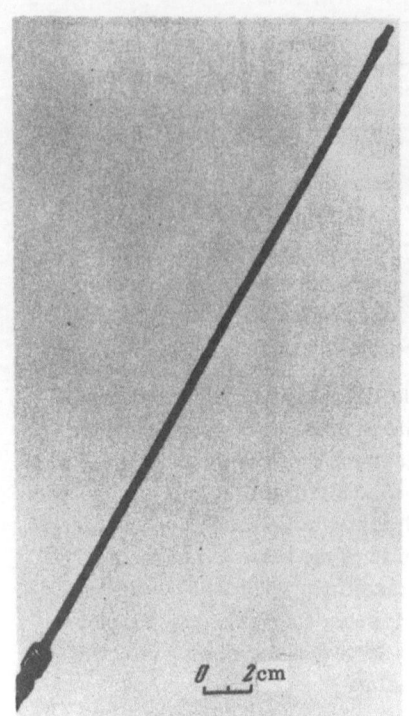

Fig. 72. Miniature cylindrical measurement sound receiver.

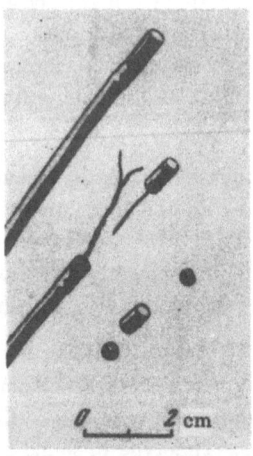

Fig. 73. Sound receiver used for medical purposes prepared in the Acoustic Institute of the Academy of Sciences of the USSR.

the element, results in a deterioration in the parameters of the sound receiver. Therefore in the case of tangentially polarized underwater sound receivers the receiver must be covered with an external insulating layer consisting of a dielectric with a low dielectric constant with a sound transmitting screen (e.g., a grid) placed outside the insulating layer.

We developed a number of different types of tangentially polarized cylindrical sound receivers for use in air and aqueous media. Since in the case of tangential polarization the use of end caps results in a reduction in sensitivity, the end surfaces of all our sound receivers of this type were screened from the effect of the sound pressure.

Figure 75 shows a typical example of this type of construction designed for use in an aqueous medium.

Sound receivers designed for use in air were usually only coated with a thin layer of vitreous enamel as a protection against humidity.

Figure 76 shows a cylindrical sound receiver for underwater application.

In a number of tangentially polarized sound receivers we used a continuous metal screen instead of a grid outside the rubber coating. However, this led to a marked reduction in sensitivity. The use of continuous screens is only advantageous at low screen thicknesses which requires development of a method of depositing a durable metallic coating directly onto the surface of the rubber or other insulating material.

The parameters of a number of investigated cylindrical sound receivers with tangentially polarized elements are given in Table 17. For comparison the theoretical sensitivities of radially polarized receivers with the same shell dimensions and fitted with end caps are also given.*

Sound receivers with 12 sections (parameters given in Table 17) could be used directly with a coaxial cable (RK-19 or RK-49) 3 to 4 m in length. The sensitivity could be increased even more by using a preamplifier located close to the sound receiver or even within the cavity of the sound receiver (at larger diameters).

The sensitivity of tangentially polarized sound receivers is highly dependent on the wall thickness of the shell (much more so than in the case of radial polarization). Thus, the wall thickness of tangentially polarized shells with diameters from 15 to 50 mm should not exceed 1 to 1.5 mm and even thinner shells are preferable. For this reason piezoelectric elements of the given type possess very thin shells and reduced mechanical strength. This is not highly important in the case

_____

* The sensitivities of radially polarized cylinders with the end surfaces screened from the effect of the sound pressure would be even smaller (by 1.5 times).

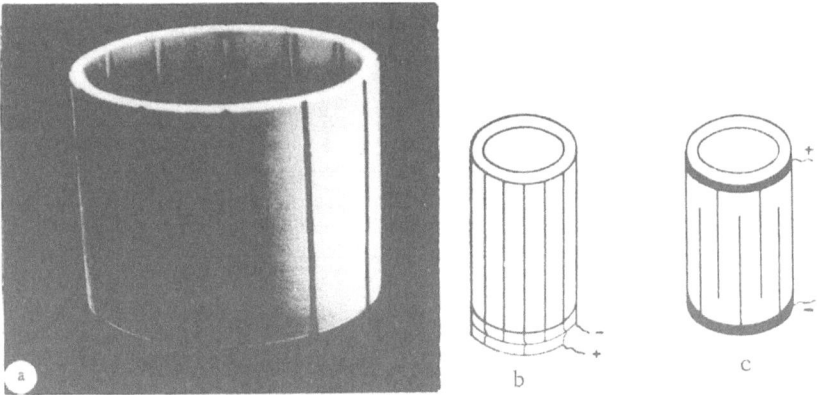

Fig. 74. a) External appearance of a tangentially polarized cylindrical element; b, c) possible methods of electrode arrangement.

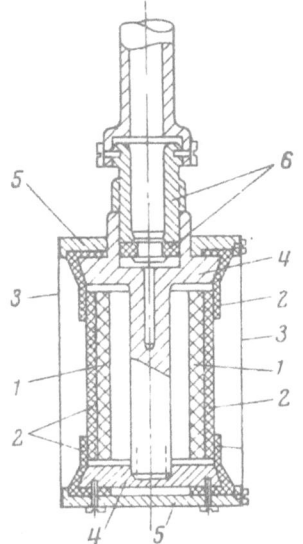

Fig. 75. Method of mounting a tangentially polarized cylindrical piezoelectric element. 1) Cylindrical piezoelectric element; 2) rubber coating; 3) external screen in the form of a grid; 4) internal metallic construction; 5) constructional members; 6) cable gland.

of sound receivers for use in air or for underwater use at low hydrostatic pressures. On the other hand in the case of underwater receivers used at considerable depths, the question of static mechanical strength is most important. However, if high sensitivity is essential, it is still possible to use tangentially polarized thin-walled shells with some means of compensating the external static pressure.

Let us now consider the directivity characteristics of the above-mentioned sound receivers. For example, Figs. 77 and 78 show the directivity characteristics in an aqueous medium (in the plane perpendicular to the axis) of sound receivers with diameters of 50 and 32 mm, respectively. Both receivers contained piezoelectric elements subdivided into 12 sections.

Comparing these characteristics with the corresponding diagrams for radially polarized cylindrical receivers, we see that in the case of tangential polarization the frequency range in which the directivity characteristic possesses a satisfactory form is narrower than in the case of radial polarization. This is undoubtedly related to the fact that in the case of radial polarization the element possesses a symmetry axis of infinite order, whereas in the case of tangential polarization the symmetry axis possesses a finite order. Therefore, diffraction phenomena should exert a much stronger effect in the case of tangential polarization than with radial polarization.

It is to be expected that disturbance in the form of the directivity characteristic should be associated with irregularities in the frequency characteristic of the sound receiver. Therefore, the actual working frequency band of a tangentially polarized receiver is somewhat narrower than in the case of radial polarization; in this case the upper frequency limit is determined not by the onset of the first resonance of the shell but by the appearance of the above-mentioned irregularities in the frequency characteristic. For example, Fig. 79 shows the frequency characteristics of tangentially polarized sound receivers with piezoelectric element diameters of 15, 30, and 50 mm. These characteristics clearly show irregularities prior to the onset of the first resonance.

TABLE 17

Cylinder dimensions, mm		Wall thickness, mm	Capacity, $\mu\mu f$	Static sensitivity, $\mu V/bar$		Sensitivity with radial polarization, $\mu V/bar$	Upper limit of working frequency bond, kc
Height	Diameter			Theoretical	Mean measured value		
40	52	1.5	500	264 -238	234	17	20
40	52	1.5	500	264 -238	250	17	20
40	52	1.5	500	264 -238	240	17	20
40	52	3.0	900	132 -110	135	16.2	20
40	52	3.0	900	132 -110	106.5	16.2	20
40	52	3.0	900	132 -110	101.5	16.2	20
40	30.9	0.9	690	127.5-115	101	10.9	50
40	32.0	1.5		73	64.4	10.7	50
23.6	14.6	1.0	—	19	—	4.75	100

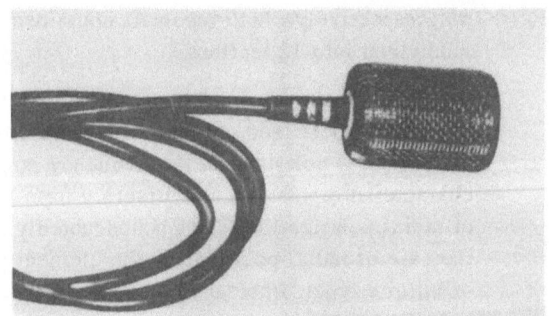

Fig. 76. Hydrophone with a cylindrical tangentially polarized ceramic element.

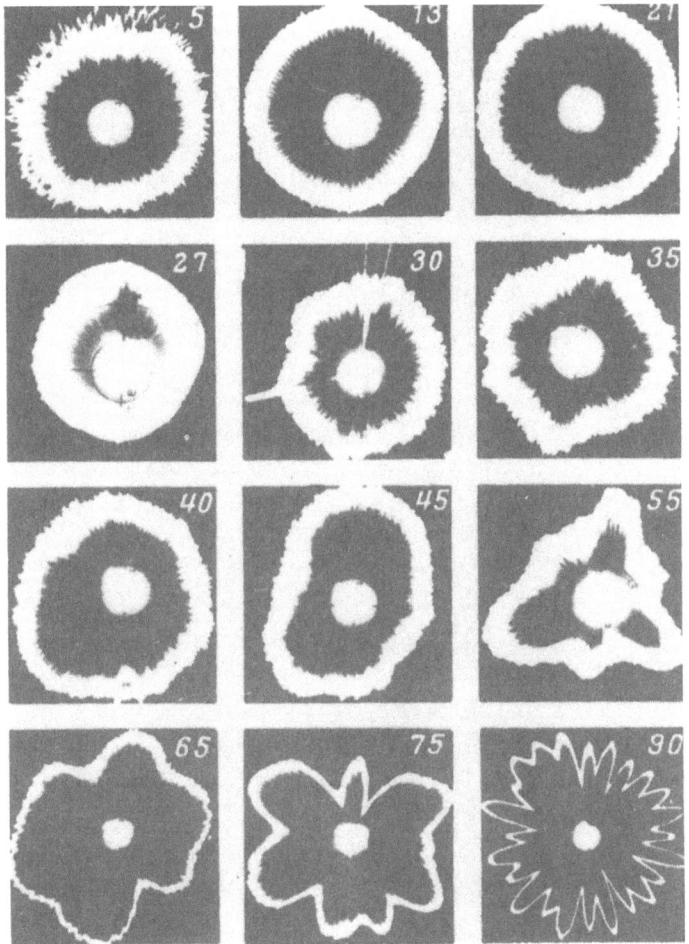

Fig. 77. Directivity characteristics of a sound receiver with a tangentially polarized cylindrical piezoelectric element recorded in the plane perpendicular to the axis. External diameter of the piezoelectric element 50 mm, n = 12. The figures give the frequencies in kc.

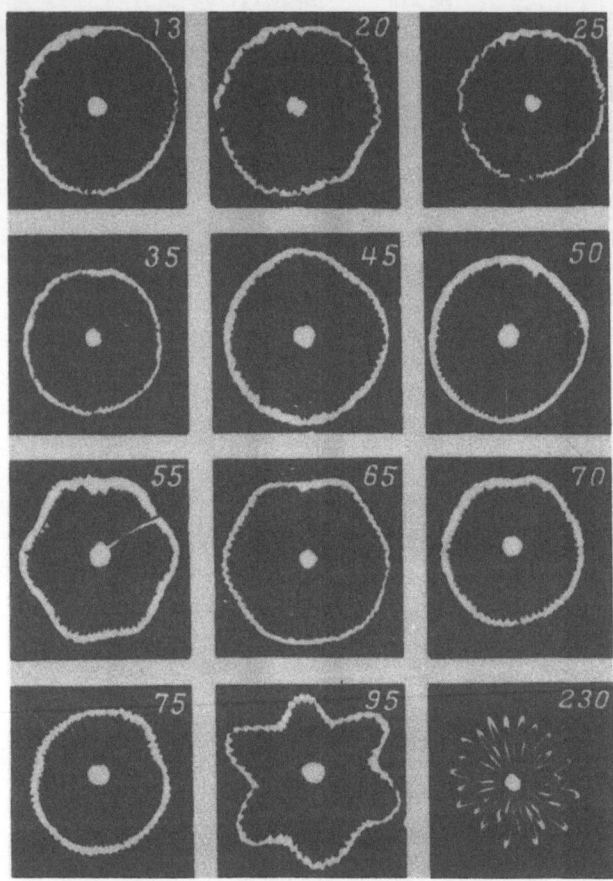

Fig. 78. Directivity characteristics of a sound receiver with a tangentially polarized piezoelectric element recorded in the plane perpendicular to the axis. External diameter of the piezoelectric element 32 mm, n = 12. The figures give the frequencies in kc.

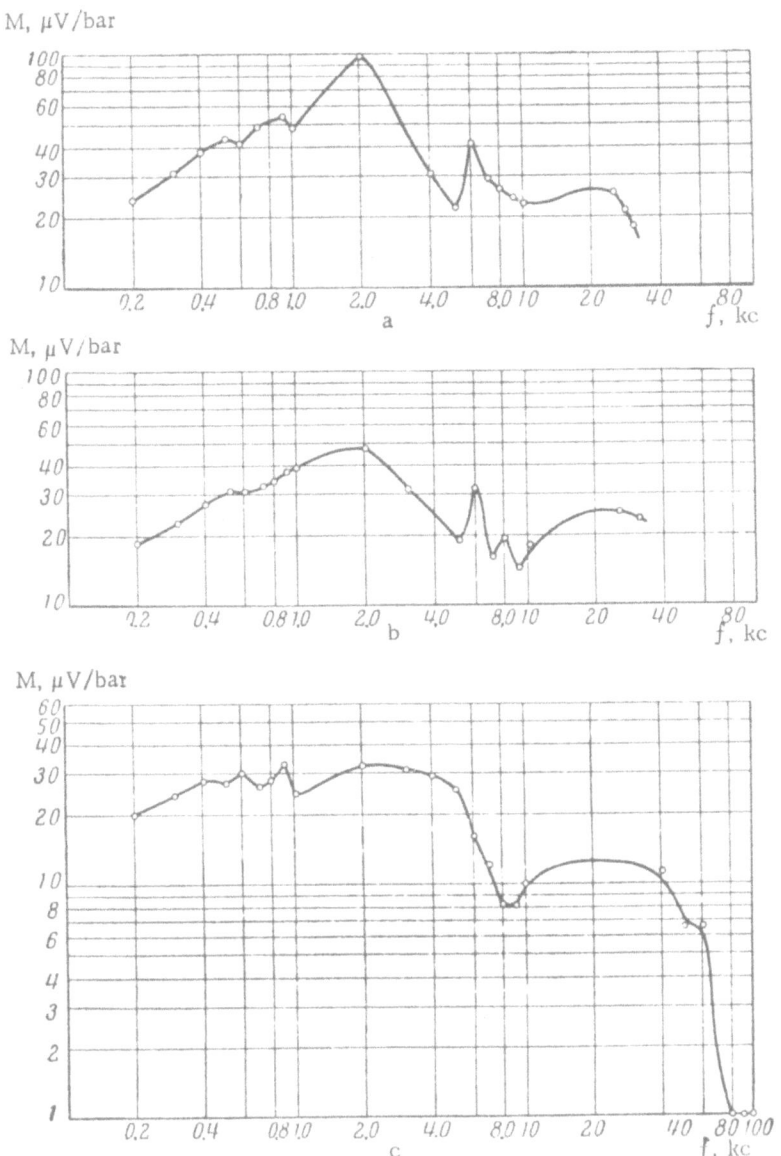

Fig. 79. Frequency characteristics of sound receivers with tangentially polarized cylindrical piezoelectric elements of barium titanate ceramic. a) External diameter of the ceramic cylinder 50 mm; b) 30 mm; c) 15 mm.

Nevertheless, it is possible to produce tangentially polarized sound receivers with adequate working frequency bands and satisfactory directivity characteristics over these frequency bands. Thus, in view of their high sensitivity, they may find even wider application than radially polarized receivers.

In conclusion we shall touch on the use of longitudinally polarized cylindrical piezoelectric elements. As follows from what was said in Section 2, for the same geometry of the ceramic shell this type of element possesses paramters intermediate between those for radially and tangentially polarized elements. However, as in the case of tangentially polarized elements they require an external insulating layer and from this point of view possess no constructional advantage over tangentially polarized receivers. For these reasons, as will be explained in detail later on, we only employed longitudinal polarization in the construction of sound receivers with plane receiving diaphragms for which this type of element is suitable.

CHAPTER IV

WIDE-BAND SOUND RECEIVERS
WITH PLANE RECEIVER DIAPHRAGMS

1. Sound Receivers of Symmetrical Construction with Two Receiver Diaphragms

We investigated a number of different types of symmetrical sound receivers with two receiver diaphragms and piezoelectric elements in the form of variously polarized cylindrical shells. This type of sound receiver is shown schematically in Fig. 80. In some cases, even when using metallic diagrams, it was possible to dispense with the insulating layer 5 (e.g., when the electrodes were not deposited over the entire surface of the piezoelectric element). Also, in some cases the diaphragms were constructed from dielectric material.

The diaphragm thickness was selected so that in the working frequency band the diaphragm could be assumed to vibrate like a piston source. The distance between the two receiver diaphragms 1, which is governed by the height of the piezoelectric element 4, was selected small in comparison with the acoustic wavelength at the upper limit of the working frequency band so that the pressures acting on the diaphragms may be considered as in-phase.

The sensitivity of this type of sound receiver is given by the simple formula

$$\frac{V}{P_0} = Kgl,$$

where l is the distance between the electrodes of the piezoelectric element, g is the corresponding piezoelectric constant, and K is the coefficient of mechanical transformation. In the given case, the coefficient of mechanical transformation is simply the ratio between the area of the diaphragm S_g, to the area of the end face of the cylindrical piezoelectric element S_n. If the radius of the diaphragm is indicated by R and the external and internal radii of the cylindrical piezoelectric element by b and a, respectively, the transformation coefficient is given by

$$K = \frac{R^2}{(b^2 - a^2)}.$$

Introducing as before the parameter $\varkappa = (b-a)/2b$, we get

$$K = \frac{1}{4}\left(\frac{R}{b}\right)\frac{1}{\varkappa(1-\varkappa)}.$$

Using this expression for the transformation coefficient we easily obtain the expression for the static sensitivity of the given type of sound receiver. For radial polarization we get

$$\frac{V}{P_0} = g_{31}b\frac{1}{2(1-\varkappa)}\left(\frac{R}{b}\right)^2,$$

or for the given values of the piezoelectric constants of the ceramic

$$\frac{V}{P_0} = 2.35\frac{b}{1-\varkappa}\left(\frac{R}{b}\right)^2.$$

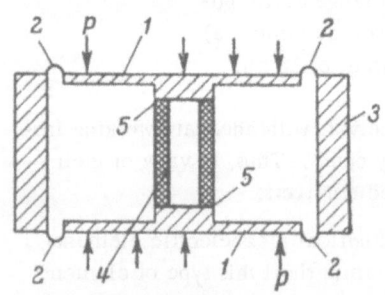

Fig. 80. Sound receivers with two receiver diaphragms. 1) Plane diaphragms; 2) diaphragm suspension rings; 3) cylindrical housing; 4) cylindrical piezoelectric element; 5) insulating layer of solid dielectric material.

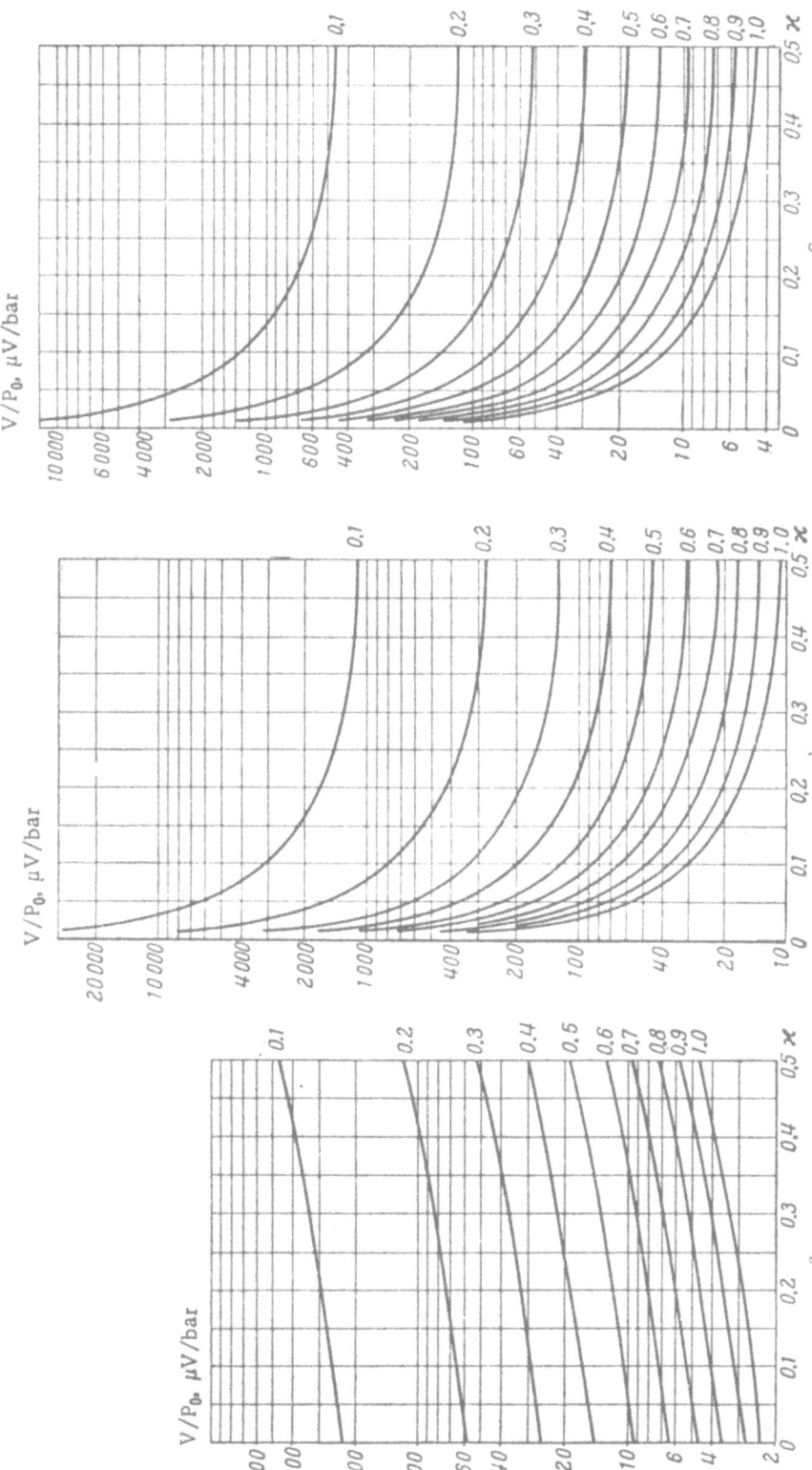

Fig. 81. Sensitivity of a sound receiver with two receiver diagrams and a piezoelectric element consisting of a tube of polarized barium titanate ceramic for different ratios of the tube and diaphragm diameters (right-hand scale). a) Radially polarized ceramic tube; b) longitudinally polarized tube; c) radially polarized ceramic tube.

Fig. 82. Sound receiver with two metal diaphragms.

TABLE 18

Sound receiver	Capacity	Sensitivity (measured) $\mu V/bar$
1	2,420	16.1
2	2,500	21.7
3	2,350	26.7

In the case of longitudinal polarization we get

$$\frac{V}{p_0} = g_{33}l \frac{1}{4\varkappa(1-\varkappa)}\left(\frac{R}{b}\right)^2 = 2.83l \frac{1}{\varkappa(1-\varkappa)}\left(\frac{R}{b}\right)^2$$

and finally for tangential polarization:

$$\frac{V}{p_0} = g_{31}l \frac{1}{4\varkappa(1-\varkappa)}\left(\frac{R}{b}\right)^2 = 1.17l \frac{1}{\varkappa(1-\varkappa)}\left(\frac{R}{b}\right)^2.$$

Thus, as might be expected the static sensitivity of a sound receiver with a plane diaphragm is proportional to a certain dimension (b or l, depending on the type of polarization) and in addition is a function of two dimensionless parameters, namely the ratio of the diameters of the diaphragm and piezoelectric element and the ratio of the wall thickness of the piezoelectric element to its external diameter.

Figure 81 gives the curves for calculating the static sensitivity for different values of R/b and $\varkappa$. For radial polarization (Fig. 81a) curves are given for b = 1 cm and for longitudinal and tangential polarization (Figs. 81b and 81c) the curves are given for l = 1 cm.

The formulae and graphs show that in sound receivers of the given type longitudinal polarization gives the best results and radial polarization the worst.

In the case of longitudinal and tangential polarization a reduction in the shell thickness leads to an increase in sensitivity but at the expense of a decrease in capacity. With the correct wall thickness from the point of view of the capacity of the piezoelectric element, radial polarization gives the best results.

Figure 82 shows a sound receiver of the given type fitted with brass diaphragms (D = 30 mm). Diaphragm mobility was achieved by means of a ring-shaped elastic element forming part of the brass diaphragm. Figure 83 shows a sound receiver with the same external diameter but with diaphragms of barium titanate ceramic. In this case the diaphragms were suspended from rubber rings.

These sound receivers employed either radially or longitudinally polarized thin-walled cylindrical piezoelectric elements. In the first case the electrodes were naturally deposited on the internal and external surfaces

Fig. 83. Sound receiver and piezoelectric element with ceramic receiver diaphragms. 1) Sound receiver housing; 2) diaphragm; 3) cylindrical piezoelectric element.

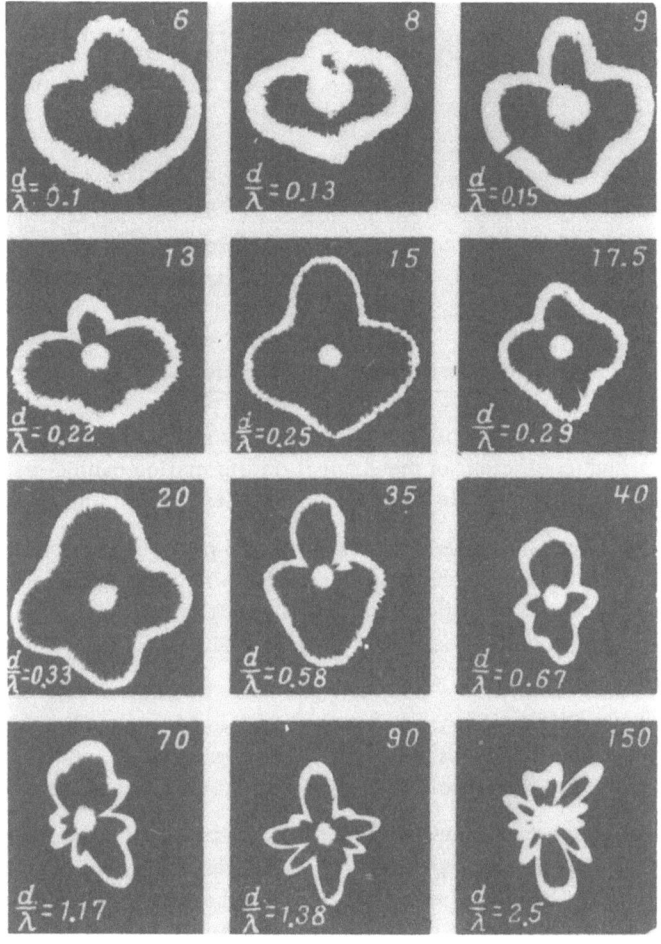

Fig. 84. Directivity characteristics for a sound receiver with two
plane diaphragms.

of the cylinder; in the second case we mostly used electrodes deposited on the external surface in the form of a double-start spiral which will be described in greater detail below [201].

Table 18 gives the principal characteristics for three specimens of the same type of sound receiver with a radially polarized piezoelectric element with a diaphragm diameter of 30 mm.

The dimensions of the ceramic piezoelectric element were as follows: external diameter $2b = 4$ mm; height $h = 12.7$ mm; wall thickness $\delta = 0.3$ mm.

Naturally, at high frequencies the directivity characteristics of such sound receivers are inferior in comparison with spherical receiver elements. Figure 84 shows the directivity characteristics for a sound receiver with a diaphragm diameter of 20 mm and a piezoelectric cylinder height of 16 mm.

Thus, at high frequencies sound receivers of this type cannot be regarded as completely nondirectional and a given orientation of the sound receiver must be maintained when using this type of receiver for measurement purposes.

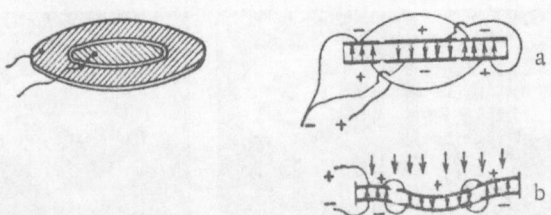

Fig. 85. Method of connecting the electrodes. a) During polarization; b) during operation of the piezoelectric element.

2. Sound Receivers with a Plane Flexible Diaphragm of Barium Titanate Ceramic

The principal of increasing the sensitivity of a sound receiver by using a piezoelectric element consisting of a deformable shell with a high coefficient of mechanical transformation can be used not only in the case of spherical or cylindrical elements but also in the case of sound receivers with a plane deformable diaphragm.

If we have a thin plane diaphragm which bends under the action of acoustic pressure, the high tangential stresses which arise in the material of such a diaphragm can be used to increase the sensitivity of the sound receiver with the aid of suitable electrodes deposited on the surface of the diaphragm and tangential polarization of the thin plane ceramic plate. The electrodes must be deposited in the form of concentric circles and the effective piezoelectric constant is g_{33}.

Polarization involving the piezoelectric constant $g_{31} = g_{32}$ can also be used with a thin flexible diaphragm of barium titanate ceramic if electrodes are deposited over the entire surface on both sides of the diaphragm. This method of polarization is similar to radial polarization of spherical or cylindrical shells.

The first method of depositing the electrodes gives high sensitivities with piezoelectric elements of low capacity. In the second method the sensitivities are not so high, but very high capacities can be obtained. Various methods can be used to support the edges of the diaphragms; the edges may be free, freely supported, or rigidly supported (clamped). In the latter case, the mechanical stresses in the central and peripheral regions of the diaphragm, arising from the pressure uniformly distributed over the diaphragm, possess opposite signs. The two regions are separated by a circular "neutral" region in which the mechanical stresses are equal to zero [202]. Therefore, when using a piezoelectric diaphragm rigidly clamped at the edge it is necessary (even in the case of thickness polarization) to use the electrode system shown in Fig. 85 instead of continuous electrodes. In addition, the method of connecting the electrodes during polarization differs from the way they are connected in the working receiver.

In the cases of freely supported edges (no restraining moment) and free edges there is no change in the sign of the mechanical stresses in the diaphragm so that continuous electrodes can be used in the case of

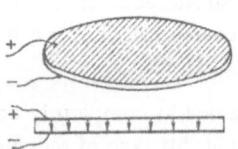

Fig. 86. Continuous electrodes for thickness polarization of a ceramic diaphragm.

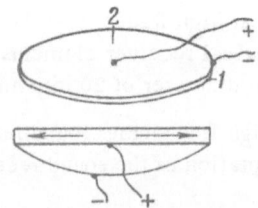

Fig. 87. Electrode system for tangential polarization of the diaphragm. 1) Ring electrode deposited on the peripheral surface of the diaphragm; 2) central electrode.

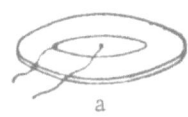

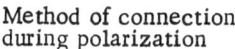
Method of connection
during polarization

b

Method of connection
during operation

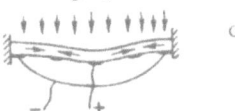
c

Fig. 88. Electrode system for opposite tangential polarization of the central and peripheral regions of the diaphragm. 1) Method of connection during polarization; 2) method of connection during operation.

thickness polarization as shown in Fig. 86. In these cases and in the case of tangential polarization the electrodes (which of course can be deposited on one side of the diaphragm only) are simple in form (Fig. 87).

In the case of tangentially polarized diaphragms with clamped edges it is not possible to use such a simple electrode system during polarization since this would result in reduced sensitivity. In this case the electrode system shown in Figs. 88a and 88b must be used resulting in different polarization directions in the central and peripheral regions of the diaphragm. During operation of the sound receiver, only the peripheral and central electrodes are used (Fig. 88c). We investigated sound receivers of this type with diaphragm diameters of 50 mm and thicknesses of 0.3 to 0.35 mm. The mean static sensitivity for thickness polarization was 40 to 50 μV/bar; the mean capacity was 60 to $70 \cdot 10^3$ $\mu\mu f$. With tangential polarization the sensitivity reached 6000 μV/bar with capacities of 10 to 20 $\mu\mu f$ [203].*

In conclusion we note that sound receivers of this type are more suitable for use in air (e.g., as microphones) in view of the brittleness of the thin diaphragms. However, we do not exclude the possibility of employing this principle in low-frequency hydrophones in cases where the external static pressure can be compensated.

---

* Tomita and Yanaguti have recently published a paper on the use of thin ceramic piezoelectric plates polarized in the direction of the plate thickness in sound receivers for use in telephony [204].

CHAPTER V

NATURAL FREQUENCY SPECTRA
OF SPHERICAL AND CYLINDRICAL SHELLS

1. Natural Frequencies of Spherical Piezoelectric Shells

In the foregoing we have investigated wide-band sound receivers for which the working frequency range was considered to be the frequency region within which the sensitivity of the receiver was approximately equal to the static sensitivity. The upper limit of this region is determined by the onset of resonance. Thus, the frequency region in which the resonance properties of the shell are important lies above the working frequency range of wide-band sound receivers and up to the present has not been examined in detail.

The sensitivity of a sound receiver at resonance frequencies is naturally much higher than the static sensitivity, so that in some cases it is advantageous to use piezoceramic sound receivers with spherical or cylindrical shells as resonators. From this point of view it is of interest to examine in detail the natural frequency spectra of spherical and cylindrical piezoelectric shells. In the case of spherical shells free from defects* we have to consider radial vibrations (zero mode) and wall-thickness vibrations.

The resonance associated with wall-thickness vibrations is a high-frequency resonance; the natural frequency region approximately coincides with the corresponding region for plane plates. However, in the case of shells we do not get one particular frequency for the first longitudinal resonance as in the case of rods, but instead a spectrum of closely adjacent resonance frequencies. On the other hand, the lower frequency radial vibrations are less complex and for small wall thicknesses we get a clearly expressed resonance at a frequency corresponding to the zero mode.

According to [205], the equation giving the natural frequencies for free radial vibration of an isotropic spherical shell of arbitrary wall thickness takes the form

$$\frac{\gamma ak + \operatorname{tg} ak \, (k^2 a^2 - \gamma)}{(a^2 k^2 - \gamma) - \gamma ak \operatorname{tg} ak} = \frac{\gamma bk + \operatorname{tg} bk \, (k^2 b^2 - \gamma)}{(b^2 k^2 - \gamma) - \gamma bk \operatorname{tg} bk}, \tag{9}$$

where b is the external radius of the shell, a is the internal radius of the shell, $k = 2\pi f/c$ is the wave number; γ is a parameter defined by the expression

$$\frac{2\lambda}{2\mu + \lambda} = 2 - \gamma,$$

where λ and μ are the Lamé coefficients.

The transcendental equation (9) can be solved graphically. In this equation the left hand side represents a function of the argument ak and the right hand side a function of the argument bk. Putting ak = z, equation (9) takes the form

$$f(z) = f\left(z \frac{b}{a}\right),$$

where b/a is a parameter. Thus, for graphical solution we must construct the principal dependence

---

*E.g., shells formed by two hemispheres joined by vitreous welding.

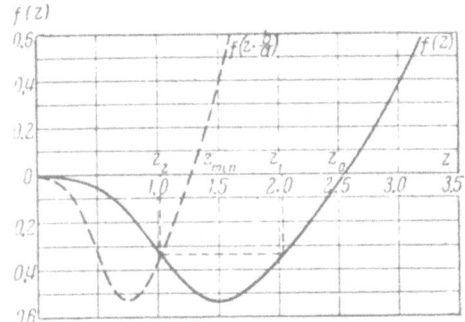

Fig. 89. The dependence $f(z)$ for $\gamma = 1.36$.

$$f(z) = \frac{\gamma z + (z^2 - \gamma)\,\mathrm{tg}\,z}{(z^2 - \gamma) - \gamma z\,\mathrm{tg}\,z}$$

for the particular value of γ for the given material. Then, assuming different values of the parameter b/a, it is a simple matter to obtain curves of $f(zb/a)$ by simply changing the scale along the horizontal axis. The intersections of the curves $f(z)$ and $f(zb/a)$ give the roots of equation (9) for the given value of γ. Since $\lambda = \dfrac{E\sigma}{(1+\sigma)(1-2\sigma)}$ and $\mu = \dfrac{E}{2(1+\sigma)}$, we can calculate the value of γ if we know the values of Young's modulus and Poisson's ratio for the material concerned. Assuming $E = 1.05 \cdot 10^{12}$ and $\sigma = 0.27$, corresponding to a sound velocity of $c = \sqrt{\dfrac{\lambda + 2\mu}{\rho}} = 4.7 \cdot 10^5$ cm/sec for a density $\rho = 5.5$ g/cm^3, we get $\gamma = 1.36$.

We note that the value of Young's modulus $E = 1.05 \cdot 10^{12}$ was assumed on the basis of measurements carried out by us on transversely polarized rods of barium titanate ceramic. According to these measurements, the sound velocity under these conditions was 4376 m/sec and exceeded the sound velocity in nonpolarized ceramic by 1.9%. This velocity was used in determining the value of Young's modulus. The value of Poisson's ratio was taken from the literature.

The continuous line curve in Fig. 89 shows the function $f(z)$ for the above-mentioned value of γ; the dotted curve is the curve of $f(zb/a)$ for $b/a = 2$. The point of intersection of these curves gives the specific value of $z = ak$ corresponding to the given value of the parameter.

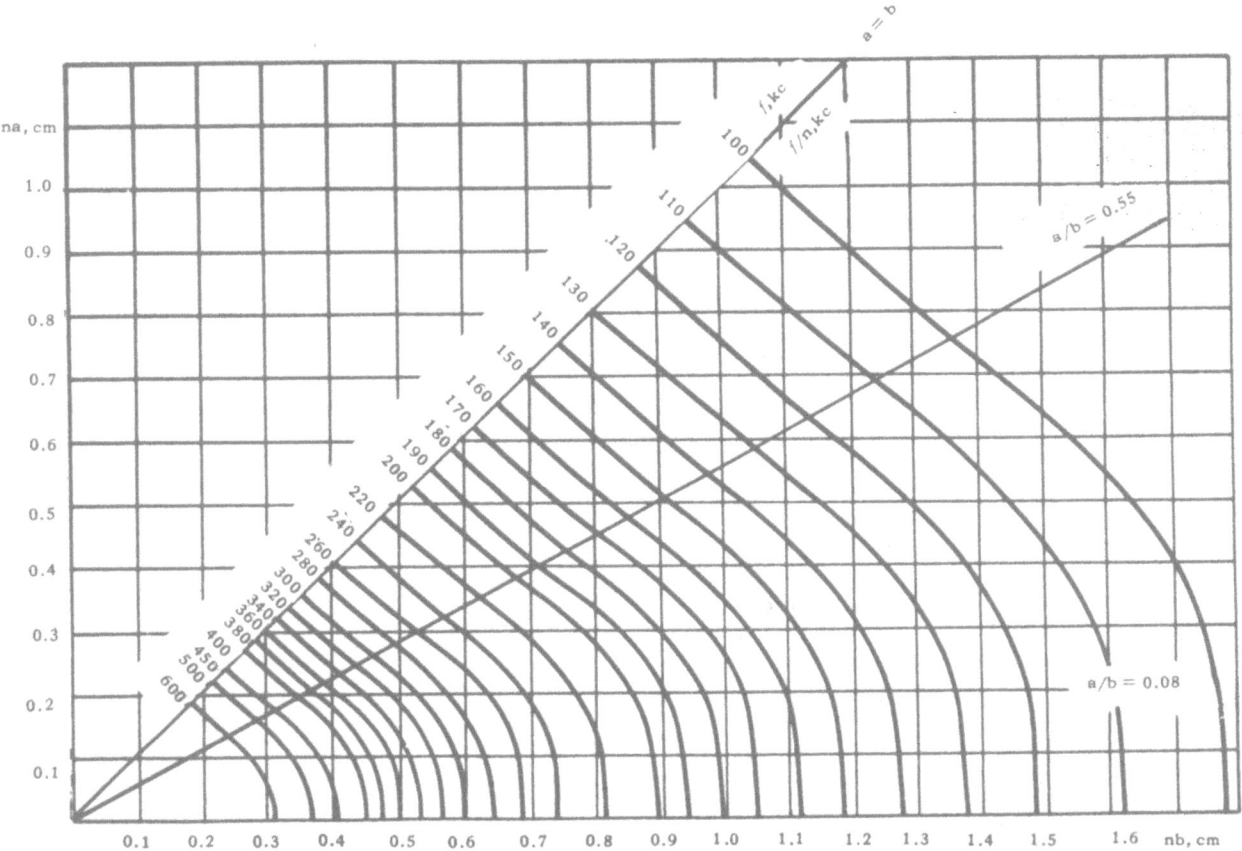

Fig. 90. Calculated resonance frequencies of radially vibrating spherical shells with arbitrary wall thickness.

TABLE 19

Specimen	External diameter, mm	Wall thickness, mm	Measured first resonance frequency f_r, kc	Calculated radial resonance frequency f_r, kc
1	10	0.8	240	230
2	10	1.0	255	235
3	8	1.0	330	300
4	16	1.0	150	135
5	5	0.5	445	445
6	30	1.5	78.5	74.0
7	30	2.0	80	75.0
8	30	3.0	81	78

With the aid of such curves and having determined the specific values of the resonance frequency f_r, it is a simple matter to construct a nomogram for determining the value of f_r for given values of a and b. Such a nomogram is shown in Fig. 90. Each point on the curve of the nomogram corresponding to a specific resonance frequency gives the combination of b and a values which results in the given resonance frequency. The straight line a/b = 1 corresponds to an infinitely thin shell; the abscissae axis, to a solid sphere. It is easily seen that in the region of the nomogram lying between the straight lines a/b = 1 and a/b = 0.55, the nomogram curves are good approximations to straight lines. In this region the shells can be considered as "infinitely thin," in which case the resultant error in determining the resonance frequency does not exceed 1%. For values of a/b in the region from zero to 0.085, the shell must be considered as a solid sphere from the point of view of calculating the natural frequency. For these special regions of the nomogram it is easy to obtain approximation formulae for the natural frequency. For $0 \leq a/b < 0.085$ and the given value of γ we get

$$f_r \approx \frac{177}{b} \text{ kc,}$$

where b is expressed in centimeters. In the "thin-shell" region, i.e., for 0.55 < a/b < 1, approximating the curves to straight lines gives

$$f_r \approx \frac{208.8}{b+a} \text{ kc.}$$

Finally, for the case of the limit thin shell (b/a = 1), we get

$$f_r = \frac{104.4}{b} \text{ kc.}$$

The approximation formulae give the resonance frequencies accurate to 1%.

We measured the natural frequencies of spherical radially polarized ceramic shells of different external diameter and wall thickness. The location of the natural frequencies was based on the frequency characteristics of the electrical impedance.

Table 19 compares the measured and calculated first resonance frequencies (for the zero mode) for spherical shells consisting of two hemispheres vitreously welded at the diametral joint. The table shows good agreement between the theoretical and experimental values. This indicates that the existence of slight defects in the actual spherical shell (diametral welding, vitreously sealed electrode aperture) has no important effect on the resonance frequencies of the shell.

88

The literature concerning the vibrations of free cylindrical shells is fairly voluminous (see for example [205]). However, up to the present there has been no complete theoretical solution of the problem of the natural frequency spectrum of cylindrical shells of finite length and arbitrary wall thickness. Therefore, we carried out a detailed experimental investigation of the natural frequency spectrum of freely vibrating cylindrical piezoelectric shells:

1) we measured the natural frequencies of a number of ceramic cylinders of equal diameter and height but differing in wall thickness; the measurements were made on radially, longitudinally, and tangentially polarized cylinders;

2) we measured the natural frequencies of cylinders of the same diameter and wall thickness but of different heights for the above-mentioned three methods of polarization.

A similar series of measurements enabled resonance associated with longitudinal vibrations (along the height of the cylinder) to be distinguished from resonances due to radial vibrations and also enabled separation of natural frequencies due to wall-thickness vibrations.

The experiments were based on the determination of the frequency dependence of the electrical impedance of each piezoelectric element. The geometric parameters of the investigated piezoelectric elements are given in Table 20 where radial, tangential, and longitudinal polarizations are indicated by R, T, and L, respectively.

Figure 91a shows a set of frequency characteristics of the electrical impedance for radially polarized cylindrical elements of the same external diameter (3.2 cm) and wall thicknesses varying from 1.3 to 0.1 cm. The curves are arranged with the wall thickness decreasing from top to bottom. The frequency in kc is plotted along the abscissae axis; the logarithmic scale along the ordinate axis is the same for all the curves. This method makes it easy to note the displacement of the natural frequencies with change in the geometrical dimensions of the shell.

Figure 91a shows that the location of the resonances associated with vibrations along the height of the cylinder (A_1, A_2, A_3) are practically independent of the wall thickness. Only the intensities of the higher order resonances (e.g., A_2 and A_3) are dependent on the wall thickness. With decrease in wall thickness, the radial resonance frequency (C) is displaced in the direction of lower frequencies. The figure clearly shows that at large wall thicknesses the radial resonance is weakly expressed in the frequency characteristic of the electrical impedance.

The location of the first resonance (B_1) of the thickness vibrations is also clearly shown in the figure; the corresponding resonance frequency is displaced in the direction of higher frequencies with decrease in wall thickness. At wall thicknesses less than 0.65 cm the structure of the frequency curve of the impedance in the region of the resonance B_1 becomes complex and instead of a single resonance frequency we get a frequency group consisting of a considerable number of closely adjacent resonant frequencies. This effect is similar to the splitting of natural frequencies observed during the vibration of piezoelectric plates [206-208]. The group of natural frequencies indicated by B_2 in Fig. 91a presumably corresponds to the second resonance of the wall thickness vibrations.

Figure 92a shows a similar set of curves for the case of longitudinal polarization for piezoelectric elements with the same external dimensions as in Fig. 91a. In this case the electrodes were deposited on the annular end surfaces of the element. The symbols A_1 and C_1 are used in the same sense as in Fig. 91a. The second longitudinal resonance A_2 is scarcely in evidence; the third longitudinal resonance can be related to the fine structure in the impedance curves here indicated by A_3 which, however, is only in evidence for small wall thicknesses. The remaining resonances are difficult to identify.

TABLE 20

	Radial polarization				Longitudinal polarization				Tangential polarization		
Specimen identification	Cylinder height, cm	External diameter, cm	Wall thickness, cm	Specimen identification	Cylinder height, cm	External diameter, cm	Wall thickness, cm	Specimen identification	Cylinder height, cm	External diameter, cm	Wall thickness, cm
1R	8	2.3	0.25	1L	8	2.3	0.25	1T	8	2.3	0.25
2R	6	2.3	0.25	2L	6	2.3	0.25	2T	6	2.3	0.25
3R	4	2.3	0.25	3L	4	2.3	0.25	3T	4	2.3	0.25
4R	3.5	2.3	0.25	4L	3.5	2.3	0.25	4T	3.5	2.3	0.25
5R	3	2.3	0.25	5L	3	2.3	0.25	5T	3	2.3	0.25
6R	2	2.3	0.25	6L	2	2.3	0.25	6T	2	2.3	0.25
7R	1.5	2.3	0.25	7L	1.5	2.3	0.25	7T	1.5	2.3	0.25
8R	1.3	2.3	0.25	8L	1.3	2.3	0.25	8T	1.3	2.3	0.25
9R	1	2.3	0.25	9L	1	2.3	0.25	9T	1	2.3	0.25
10R	0.7	2.3	0.25	10L	0.7	2.3	0.25	10T	0.7	2.3	0.25
11R	0.5	2.3	0.25	11L	0.5	2.3	0.25	11T	0.5	2.3	0.25
12R	1.5	3.2	1.3	12L	1.5	3.2	1.3	12T	1.5	3.2	1.3
13R	1.5	3.2	1.1	13L	1.5	3.2	1.1	13T	1.5	3.2	1.1
14R	1.5	3.2	1.05	14L	1.5	3.2	1.05	14T	1.5	3.2	0.05
15R	1.5	3.2	1	15L	1.5	3.2	1	15T	1.5	3.2	1
16R	1.5	3.2	0.95	16L	1.5	3.2	0.95	16T	1.5	3.2	0.95
17R	1.5	3.2	0.9	17L	1.5	3.2	0.9	17T	1.5	3.2	0.9
18R	1.5	3.2	0.8	18L	1.5	3.2	0.8	18T	1.5	3.2	0.8
19R	1.5	3.2	0.65	19L	1.5	3.2	0.65	19T	1.5	3.2	0.65
20R	1.5	3.2	0.5	20L	1.5	3.2	0.5	20T	1.5	3.2	0.5
21R	1.5	3.2	0.33	21L	1.5	3.2	0.33	21T	1.5	3.2	0.33
22R	1.5	3.2	0.25	22L	1.5	3.2	0.25	22T	1.5	3.2	0.25
23R	1.5	3.2	0.15	23L	1.5	3.2	0.15	23T	1.5	3.2	0.15
24R	1.5	3.2	0.1	24L	1.5	3.2	0.1	24T	1.5	3.2	0.1

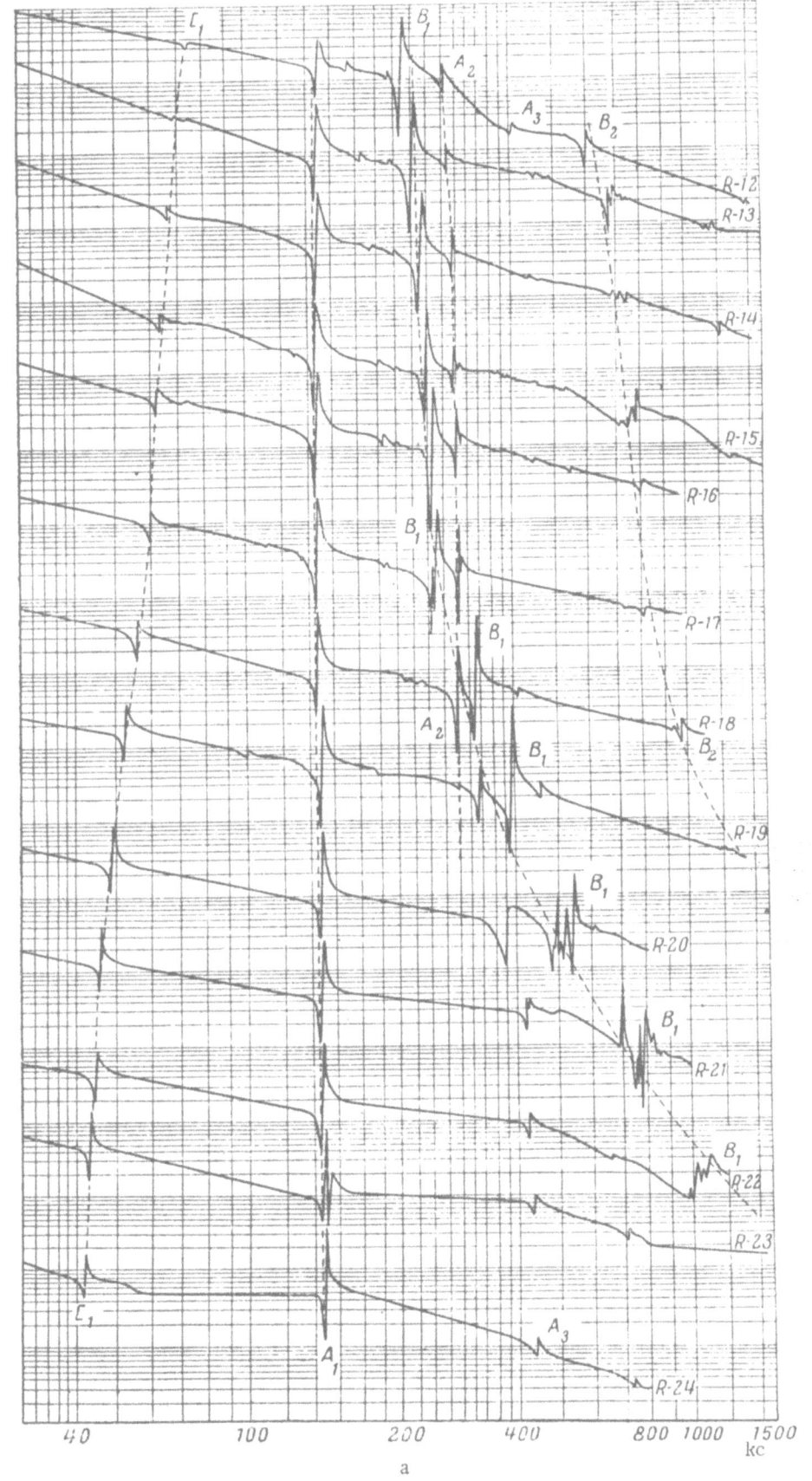

Fig. 91. Frequency dependence of (a) the electrical impedance and (b) the phase angle between the current and voltage in radially polarized cylinders of barium titanate ceramic with different wall thicknesses. h = const = 1.5 cm.

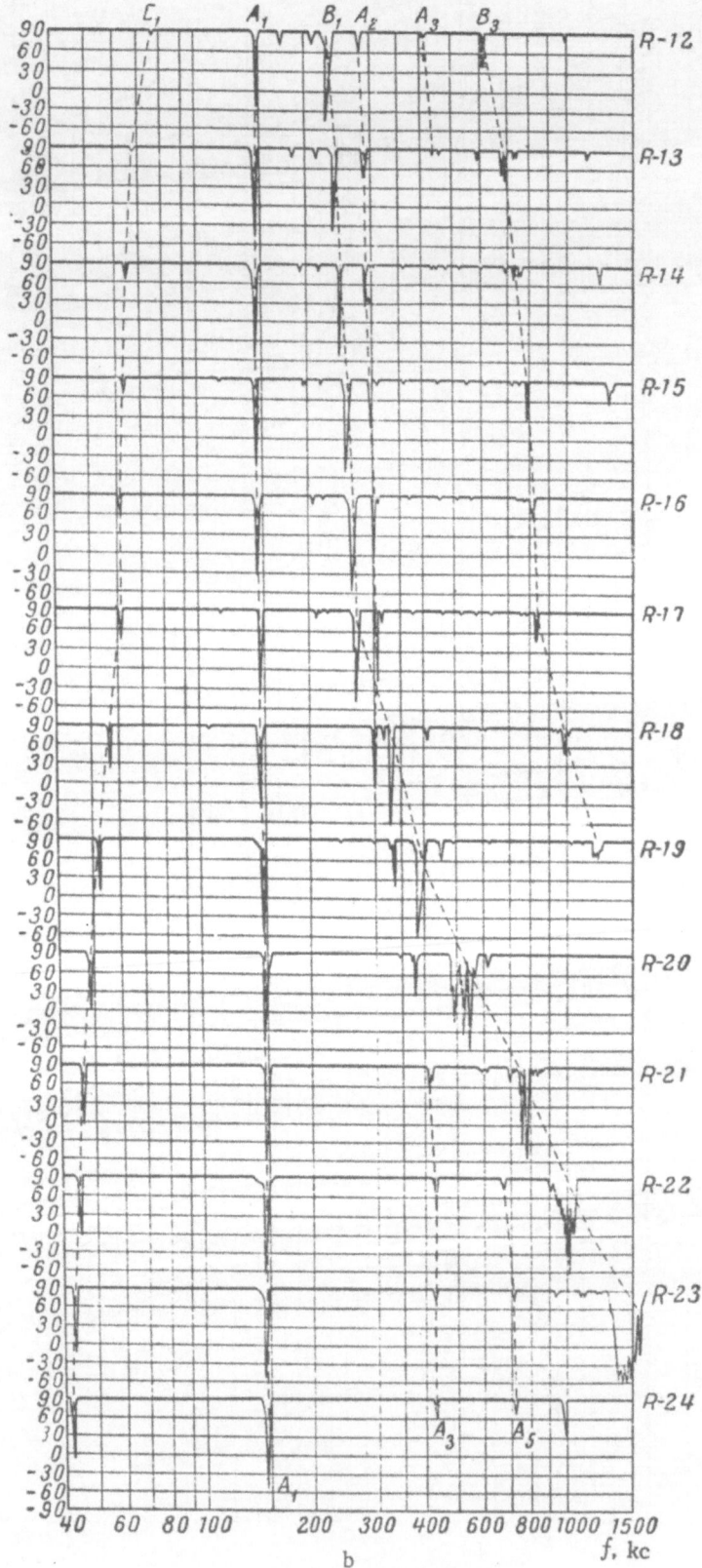

Fig. 91. (Continued)

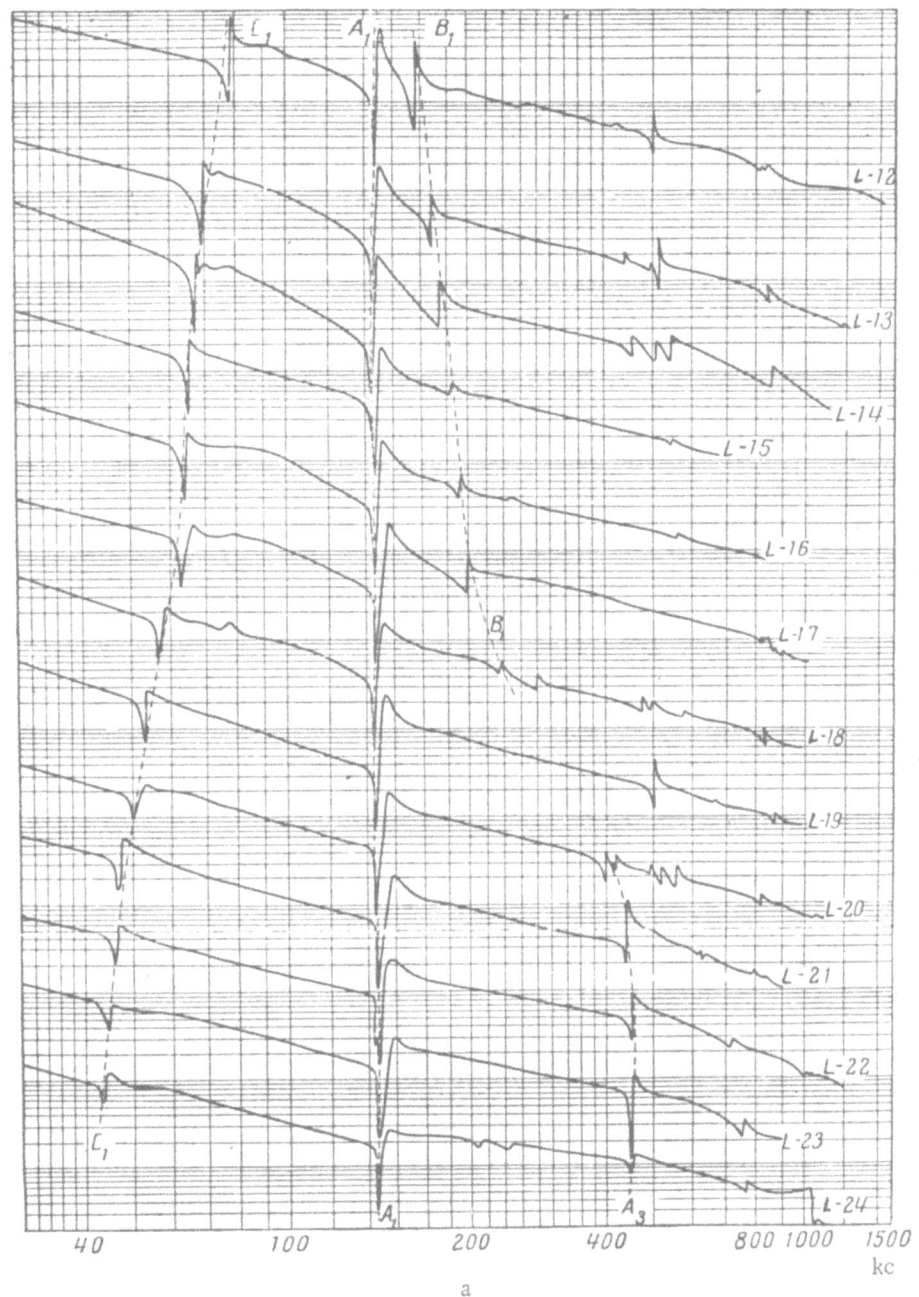

Fig. 92. Frequency dependence of (a) the electrical impedance and (b) the phase angle between the current and voltage in longitudinally polarized cylinders of barium titanate ceramic with different wall thicknesses. h = const = 1.5 cm.

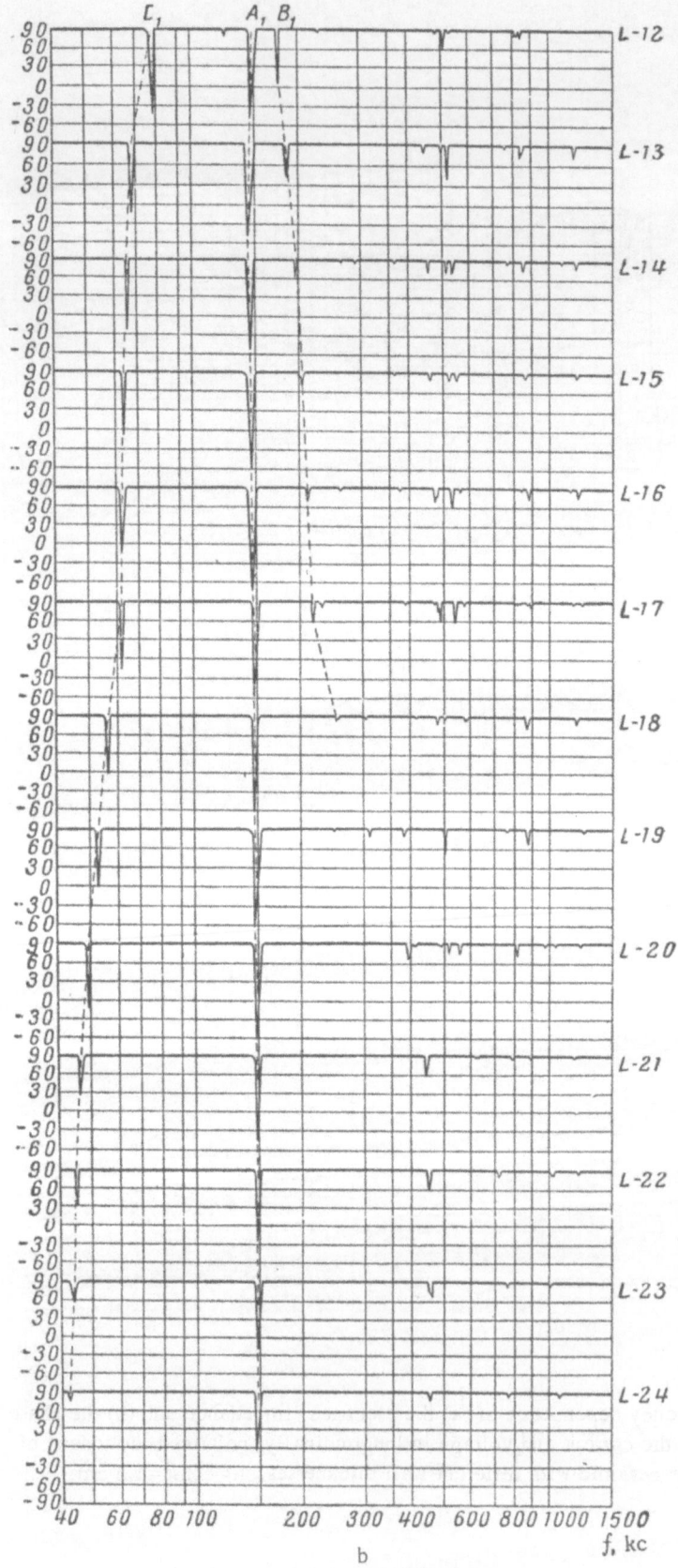

Fig. 92. (Continued)

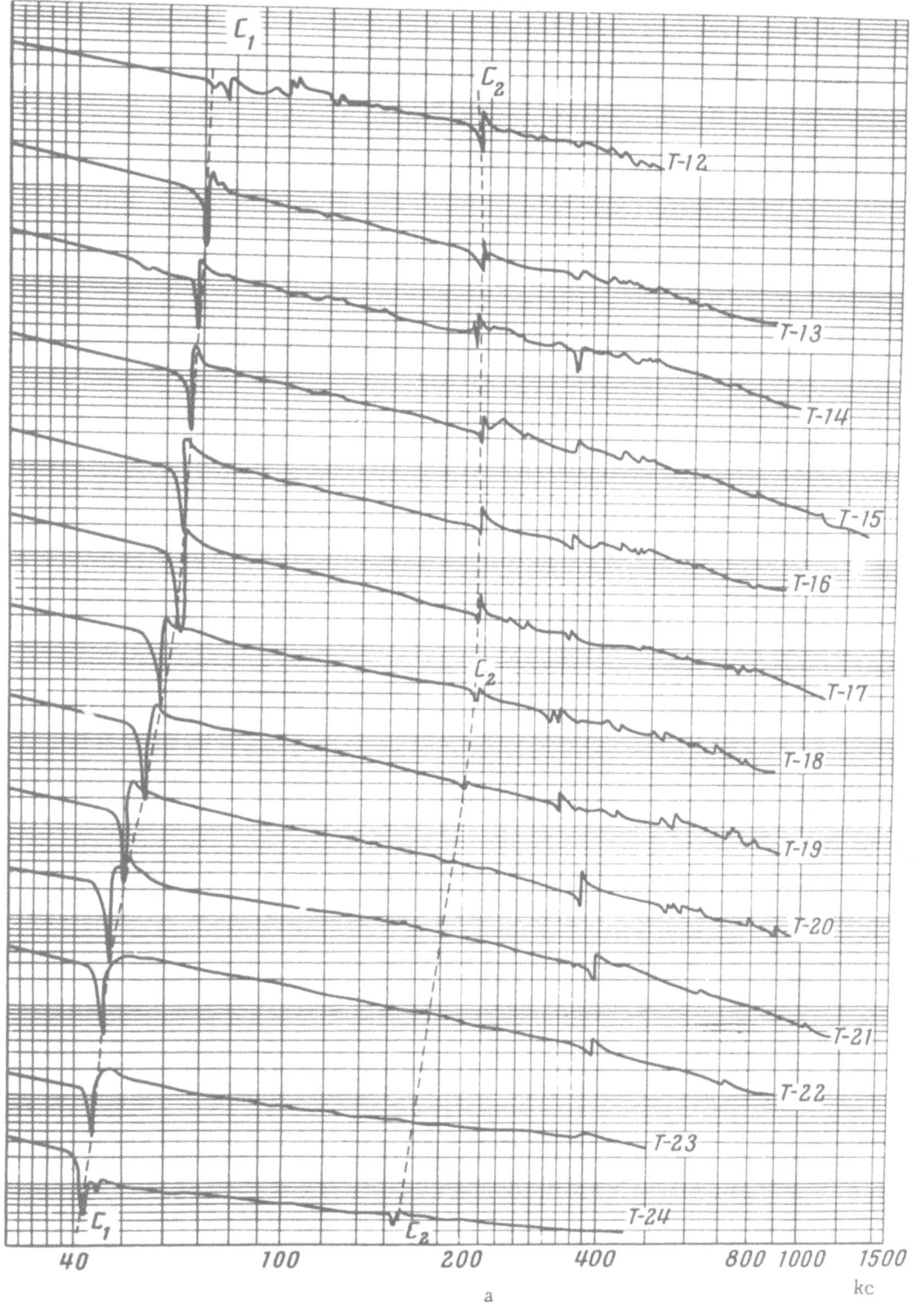

Fig. 93. Frequency dependence of (a) the electrical impedance and (b) the phase angle between the current and voltage in tangentially polarized cylinders of barium titanate ceramic with different wall thicknesses. h = const = 1.5 cm.

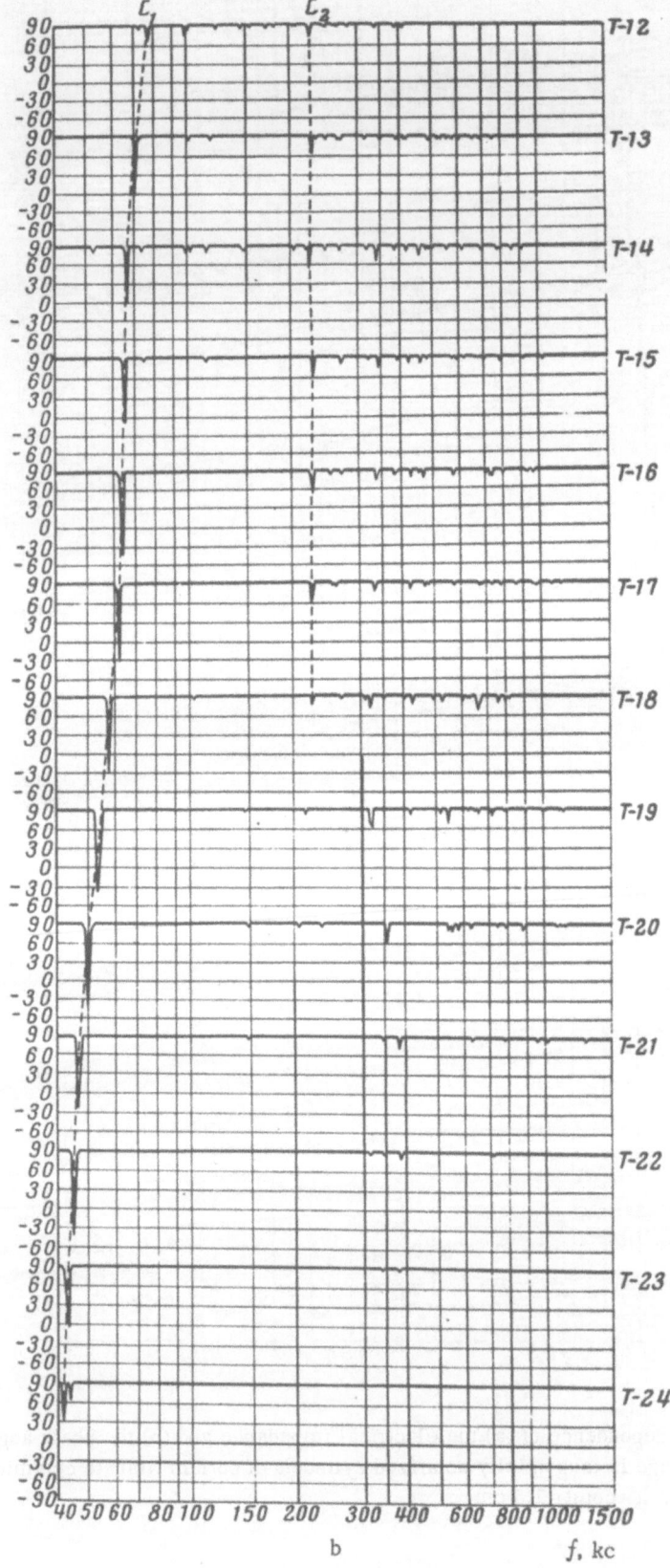

Fig. 93. (Continued)

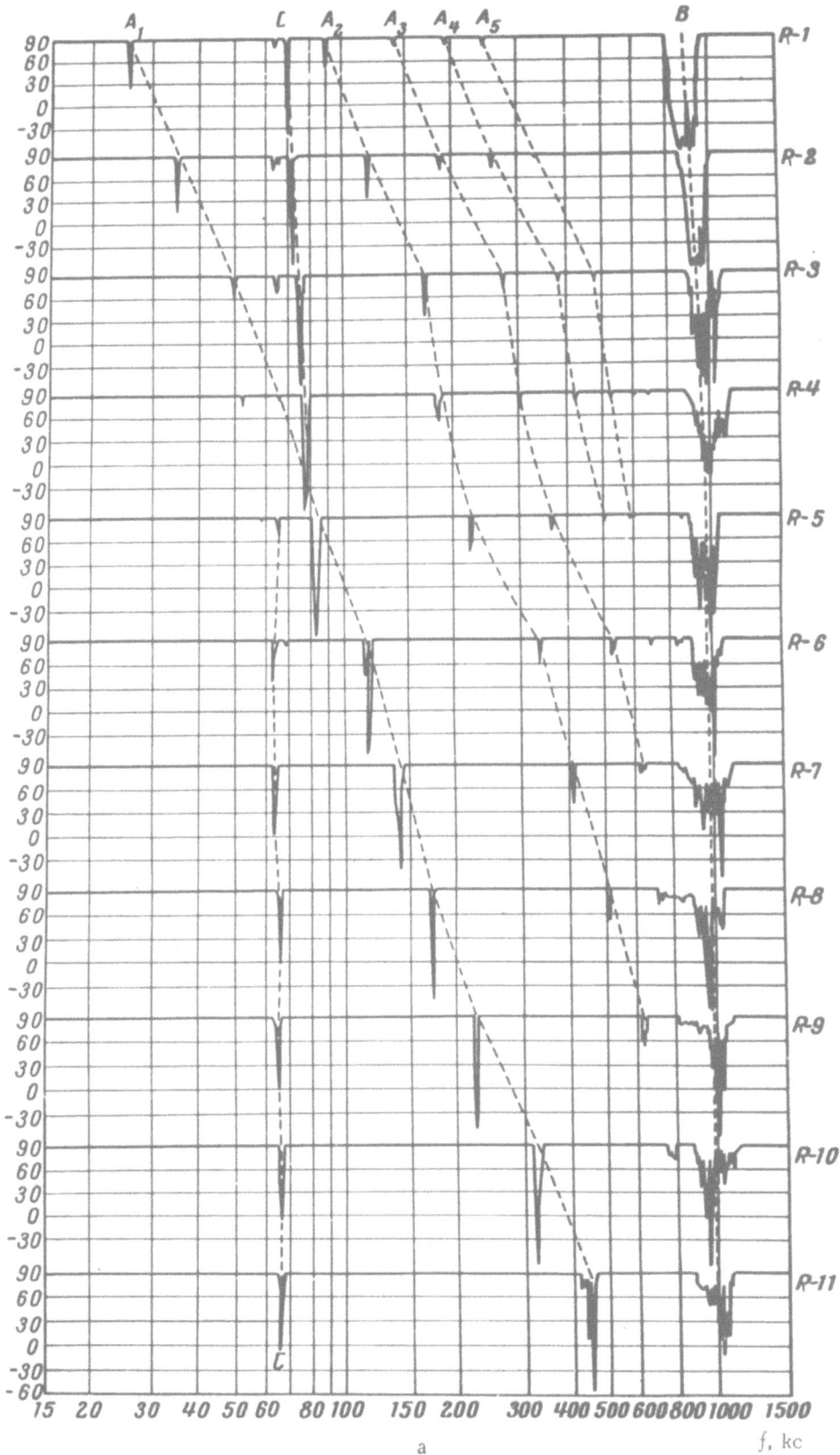

Fig. 94. Natural frequency spectra for cylindrical tubes of barium titanate ceramic of different axial length. a) Radially polarized; b) longitudinally polarized; c) tangentially polarized.

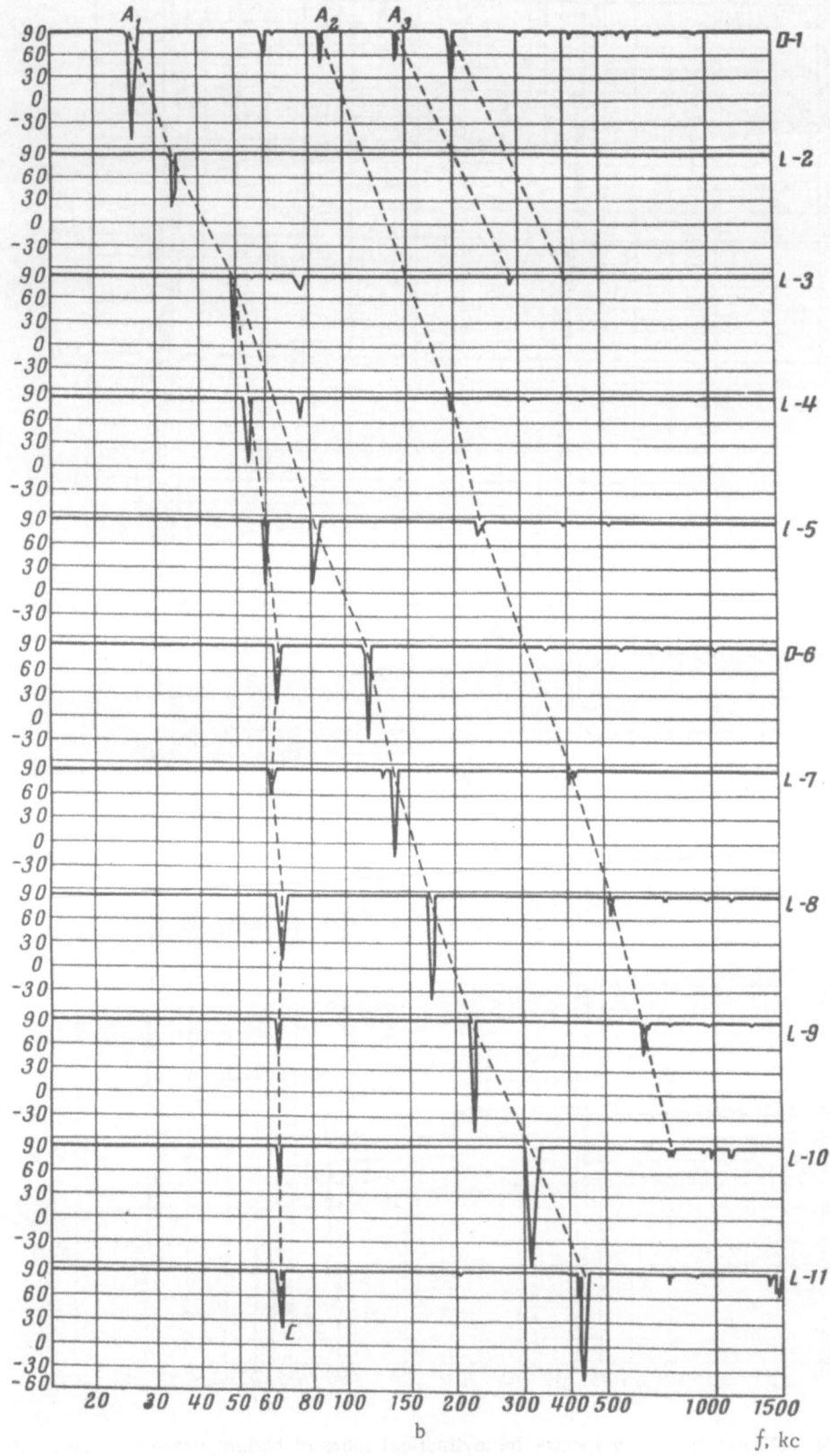

Fig. 94. (Continued)

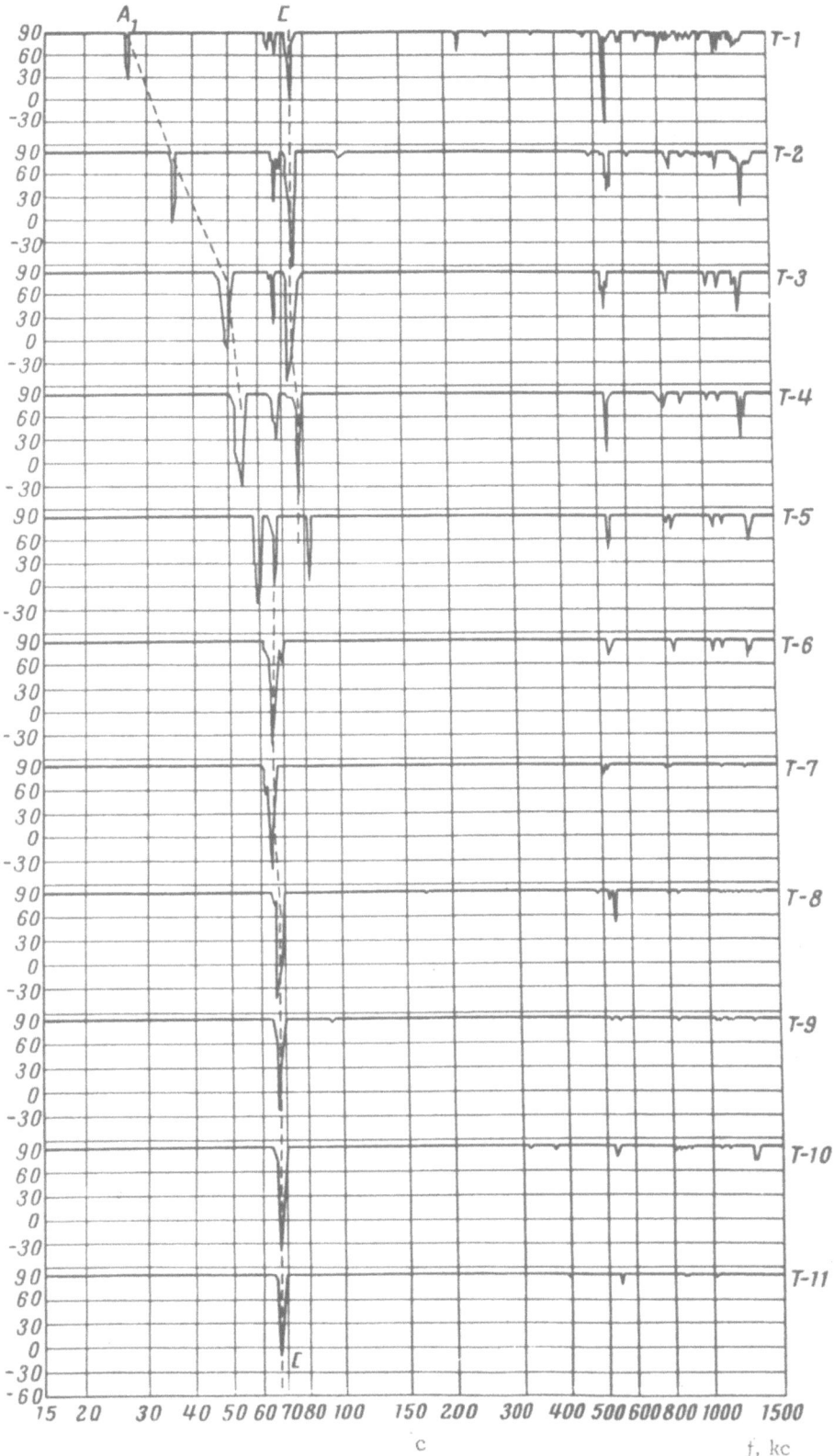

Fig. 94. (Continued)

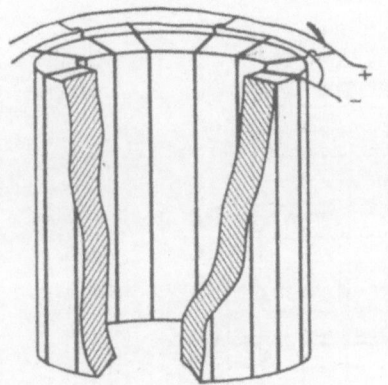

Fig. 95. Method of depositing the electrodes on the surface of a whole ceramic cylinder.

Figure 93 shows the corresponding curves for the case of tangential polarization for piezoelectric elements of the same dimensions. The elements were subdivided into 12 sections. The electrodes were deposited on the surface of the whole ceramic cylinder as shown in Fig. 95, i.e., each electrode was deposited along opposite internal and external generating lines which were connected radially at the end surfaces.

In this case, the radially resonance C_1 is very clearly expressed; we may assume that the fine structure indicated by C_2 in Fig. 93a is related to the second radial resonance. The longitudinal and thickness resonances are here practically absent.

The following remarks apply generally to Figs. 91a, 92a, and 93a. Firstly, the most clearly expressed modes of vibration are those for which the direction of deformation coincides with the direction of polarization; these vibrations are very clearly expressed on the electrical side (with the exception of certain cases of electrode connection). A good example of this is provided by Fig. 93a where tangential polarization results in clearly expressed radial vibrations corresponding to extremely high tangential mechanical stresses. The remaining types of vibration (longitudinal and thickness) are only weakly expressed since they are due solely to the existence of Poisson's ratio.

The case of longitudinal polarization shown in Fig. 92a appears to be intermediate between the cases of radial and tangential polarization. In this case the resonances in the longitudinal vibrations which coincide with the polarization direction are most clearly in evidence. The radial (C) and thickness (B_1) resonances are more weakly expressed since the corresponding vibrations are due only to the existence of Poisson's ratio and in these cases the direction of deformation does not coincide with the direction of polarization.

Finally, in the case of radial polarization, the thickness resonances are naturally most clearly in evidence since in this case the direction of deformation coincides with the direction of polarization. The gradual convergence between the longitudinal (A_1) and thickness (B_1) resonances evident in Fig. 91a applies only within the limits of the given experiment since for specimens with the number 18 the height and wall-thickness dimensions converge.

Finally, we note the "disappearance" of certain types of resonance observed in the figures. This degeneration observed on the electrical side is due to the fact that for a given electrode arrangement and given polarization and deformation directions a complex mechanical stress pattern may arise in the region enclosed by the electrodes characterized by the presence of a zone in which the stresses are of opposite sign. This leads to the known type of charge compensation and a decrease in the coupling coefficient.

Our conclusions are confirmed by measurements of the frequency dependence of the phase angle φ (phase displacement between the current and voltage) carried out on the same specimens. The corresponding frequency dependences of the phase angle are given in Figs. 91b, 92b, and 93b for comparison with the frequency dependences of the electrical impedance.

We carried out a second series of measurements on specimens of different height, the remaining dimensions being unchanged. These measurements were made on piezoelectric cylinders Nos. 1 to 11 of Table 20; in all cases the wall thickness was 0.25 cm, the diameter 2.3 cm, i.e., the measurements were carried out on thin-walled piezoelectric elements. The measurements were carried out by determining the frequency characteristics of the phase angle of the electrical impedance. At frequencies fairly well removed from the resonance frequencies the phase angle φ did not differ greatly from 90° since the electrical impedance of the transducer is practically a purely reactive capacitive resistance. Deviation of the angle φ from 90° is caused only by the internal dielectric losses in the ceramic. At the resonance frequencies the angle φ suddenly changes its sign and the higher the Q-value the closer the angle φ approaches to −90°.

TABLE 21

Relationship between D and l	Polarization	Longitudinal resonance (along the cylinder height)	Radial resonance	Wall thickness resonance
$l > D$	Radial	+	+	+
	Longitudinal	+	−	−
	Tangential	+	+	−
$D > l$	Radial	+	+	+
	Longitudinal	+	+	−
	Tangential	−	+	−

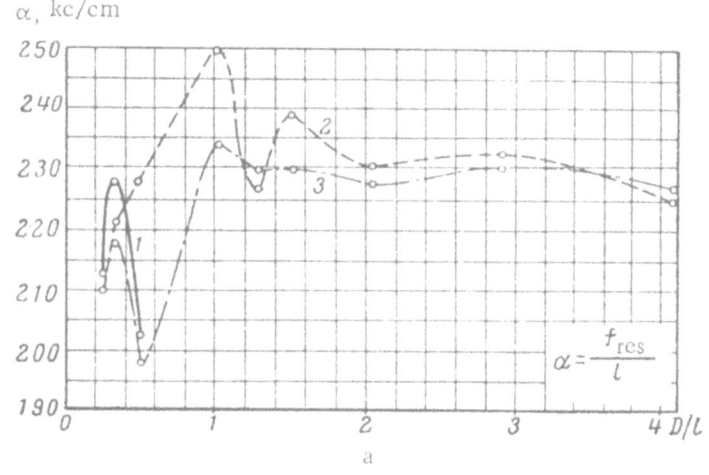

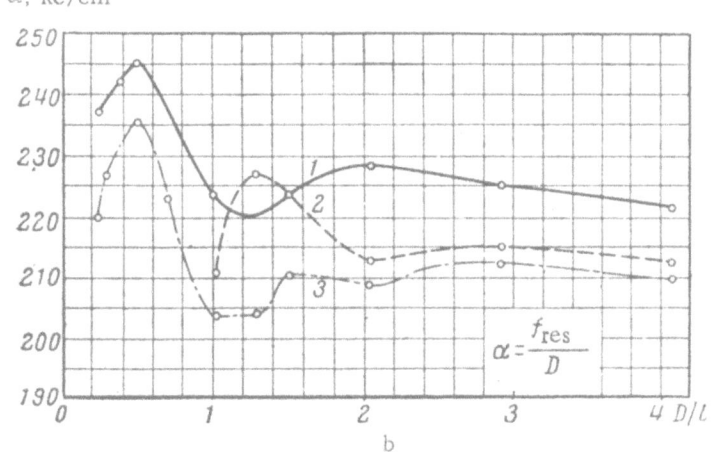

Fig. 96. Frequency constant for vibrations of a cylinder of barium titanate ceramic for different ratios of the diameter of the cylinder to its axial length. a) For longitudinal vibrations; b) for radial vibrations.

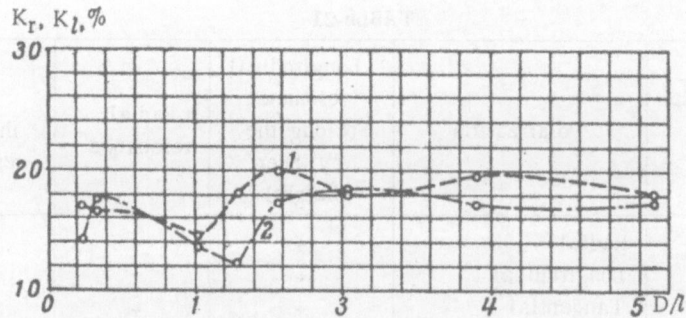

Fig. 97. Electromechanical coupling coefficient for longitudinal and radial vibrations of radially polarized barium titanate ceramic cylinders with different diameter to axial length ratios.

Thus, the resonances appear as sharp dips in the frequency characteristics of the phase angle with the extent of the dip increasing with the Q-value at the given resonance frequency.

Figure 94 shows the variation in the frequency characteristics of the phase angle with change in the height of the cylindrical piezoelectric element for the cases of radial, longitudinal and tangential polarization. In all the diagrams the cylinder height decreases from top to bottom.

Figure 94a which shows the spectra for the case of radial polarization clearly shows the oblique traces of A_1, A_2, A_3, A_4 ..., corresponding to the first, second, third, fourth, etc., longitudinal resonances. The radial resonance C is also clearly expressed. At low wall thicknesses, the thickness resonances occur at higher frequencies, and although clearly in evidence they show a considerable degree of splitting.

In the case of longitudinal polarization (Fig. 94b) the longitudinal resonances A_1, A_2, A_3 are also clearly expressed. The radial resonance C is clearly in evidence at small cylinder heights but disappear at larger heights for the reasons mentioned above; in the given case it vanishes due to the creation along the cylinder height of zones in which the deformations are of opposite sign. Thickness resonance is practically absent in this case.

In Fig. 94c (tangential polarization) the radial resonances are naturally most clearly in evidence. The first longitudinal resonance A_1 is fairly well expressed at large cylinder heights. At small cylinder heights even the first longitudinal resonance does not appear for the reasons given above for the case of longitudinal polarization. The remaining resonances are weakly expressed and are difficult to identify.

Thus, the experiments involving change in height of thin-walled cylinders confirm the data obtained from measurements carried out on thick-walled cylinders.

In fact, however, during measurement of the frequency dependence of the input electrical parameters (electrical impedance $|z|$ and phase angle φ) not every type of natural resonance is indicated by measurements on the electrical side for different types of polarization of cylindrical specimens with the above-mentioned very simple systems of uniformly distributed electrodes. Thus, in the case of longitudinally polarized tubes which are long in comparison to their diameter, radial resonance is not observed; in the case of tangentially polarized tubes which are short in comparison with their diameters, longitudinal resonance is not observed. Table 21 shows the resonances which are clearly expresed on the electrical side for the three possible types of polarization. The "+" sign indicates that the particular type of resonance is clearly observed experimentally and the "−" sign indicates that the resonance is completely absent or only very weakly in evidence. The decisive factor appears to be the relationship between the diameter D and the axial length (height) l of the cylinder.

From everything which has been said above we may conclude that for a given type of polarization, in order to obtain a satisfactory coupling coefficient at a given mode of vibration, it is necessary to locate a requisite number of electrodes in accordance with the distribution in the body of the piezoelectric element of the dynamic stresses corresponding to the given vibration mode.

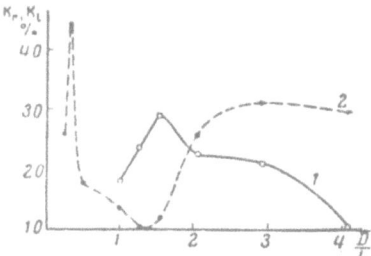

Fig. 98. Electromechanical coupling coefficient for (1) radial and (2) longitudinal
vibrations of longitudinally polarized barium
titanate ceramic cylinders with different
diameter to axial length ratios.

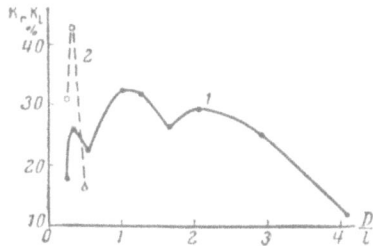

Fig. 99. Electromechanical coupling coefficient for (1) radial and
(2) longitudinal vibrations of tangentially polarized cylinders with different diameter to axial length ratios.

In determining the overall dimensions of cylindrical ceramic piezoelectric elements for sound receivers
it is advantageous to know the frequency constants of the ceramic for longitudinal and radial vibrations of the
cylinder for different types of polarization. In analogy with the case of a plane ceramic plate we define the
frequency constant as the ratio of the resonance frequency to the height or external diameter of the cylinder.

On the basis of previously determined frequency characteristics of the modulus and phase of the electrical impedance for cylindrical tubes of different geometric dimensions, we constructed the dependence of
the frequency constants for radial and longitudinal vibrations of a thin-walled ceramic cylinder on the ratio
of the external diameter of the cylinder D to its axial length l. This dependence is shown in Fig. 96a for the
case of longitudinal vibration for (1) tangential, (2) longitudinal, and (3) radial polarization. Figure 96b shows
the corresponding dependences for the case of radial vibrations.

The frequency constants differ for different types of polarization of the cylindrical piezoelectric elements and are practically independent of the ratio D/l for flat cylinders with D/l > 2, but in the case of cylinders for which the diameter and height were commensurable we obtained frequency constants differing from
those for flat cylinders.

We know [209] that the electromechanical coupling coefficient can be determined from the resonance
and antiresonance frequencies for radial and longitudinal vibrations of cylindrical piezoelectric elements. The
electromechanical coupling coefficient for radial vibrations is given by

$$k_r = \sqrt{\frac{2\pi\Delta f_2}{f_r}}.$$

The electromechanical coupling coefficient for longitudinal vibrations (resonance along the cylinder
height) is given by

$$k_l = \sqrt{\frac{\pi^2}{8}\frac{2\Delta f_l}{f_l}} = \frac{d_{ik}}{\sqrt{\frac{\varepsilon}{4\pi}\delta}},$$

where Δf_r and Δf_l are the differences between the resonance and antiresonance frequencies for radial and
longitudinal vibrations, respectively; f_r and f_l are the corresponding antiresonance frequencies; δ is the compliance constant; d_{ik} is the piezoelectric modulus.

Figures 97, 98, and 99 show the values of the electromechanical coupling coefficients obtained for longitudinal and radial vibrations with variously polarized thin-walled ceramic cylinders in dependence on the
ratio of the diameter of the cylinder D to its axial length l. Figure 97 shows that radially polarized cylinders
possess an approximately constant value of the electromechanical coupling coefficient for (1) longitudinal
and (2) radial vibrations (approximately 17 to 18%) independent of the ratio D/l.

Figure 98 shows the results for longitudinally polarized cylinders for (1) radial and (2) longitudinal resonances. In this case the electromechanical coupling coefficients are different for radial and longitudinal vibrations. This is to be expected since in the case of longitudinal resonance and longitudinal polarization the coupling coefficient is determined by the piezoelectric constant d_{33} whereas in the case of radial resonance it is related to the constant d_{31}.

Figure 99 shows the corresponding results for tangential polarization.

The data given must be regarded as only approximate. They indicate the magnitude of the change in the electromechanical coupling coefficient with change in the diameter to axial length ratio for different types of polarization. With the aid of these data it is possible to estimate the limits of variation of D/l within which the electromechanical coupling coefficient will possess a sufficiently large value.

It must be borne in mind that the data obtained (frequency constants and electromechanical coupling coefficients) relate firstly only to the case of the simplest systems of electrodes uniformly distributed over the curved surface of the cylinder and secondly to the case of whole unsectioned cylinders.

3. Use of Sound Receivers with Cylindrical Piezoceramic Elements as Resonance Receivers

It follows from what has been said that sound receivers with cylindrical piezoelectric elements of barium titanate ceramic possess a spectrum of natural frequencies containing a number of pronounced resonances. The sensitivity of the sound receiver at these resonances considerably exceeds the static sensitivity. In certain cases this enables such receivers to be used as resonance receivers for the reception of sound of specific frequencies. Thus, in the case of radially polarized cylindrical sound receivers at high ultrasonic frequencies it is possible to use the region of wall-thickness resonances. Of course, it is also possible to use resonances of lower frequency, especially in the case of measurement sound receivers. For example, Table 22 gives the sensitivity of radially polarized cylindrical sound receivers at the first radial resonance.

It is of course obvious that sound receivers with spherical piezoelectric elements can also be used as resonance receivers; the sensitivity of such elements at the frequency of the first radial resonance with a shell thickness to diameter ratio $\varkappa \approx 0.1$ is 1.5 to 3 times greater than the static sensitivity.

The directivity characteristics of radially polarized sound receivers in the low-frequency resonance region is very close to circular. In the case of tangentially polarized cylinders, certain resonance frequencies may be associated with marked deviation of the directivity characteristic from circular; however, generally speaking such deviations do not exceed 6 db.

Therefore, both spherical and cylindrical sound receivers can be successfully employed as nondirectional receivers in the low-frequency resonance region with sensitivities somewhat higher than the static sensitivity.

TABLE 22

Diameter of piezoelectric element, mm	Capacity of sound receiver, $\mu\mu f$	Sensitivity at resonance, $\mu V/bar$	Static sensitivity, $\mu V/bar$
6	2,200	2.1	1.4
15	4,000	4.5	3.76
30	8,000	10	7.0
50	2,000	15	11.6

CHAPTER VI

RESONANCE SOUND RECEIVERS
WITH CERAMIC PIEZOELECTRIC ELEMENTS

1. Resonance Sound Receivers with a Plane Receiver Diaphragm and Cylindrical Piezoelectric Elements

We investigated a number of types of resonance sound receiver with transducer elements of barium titan-ate ceramic. We concerned ourselves chiefly with piezoelectric elements in the form of longitudinally or ra-dially polarized cylinders connected in some manner to a receiver diaphragm.

Transducers of this type with different resonance frequencies were used for example as sound radiators for creating sound fields in underwater sound instrumentation and for measurements in an underwater sound tank (for calibration by the reciprocity method with the transducer acting as a receiver-transmitter or as an auxiliary transmitter, etc.).

This type of resonance transducer with a cylindrical piezoelectric element 1 is shown in Fig. 100; part of the casing is cut away to show the internal construction.

The rubber sealing ring grips the outer edge of the diaphragm; on the front side the diaphragm is exposed and acts as the sound pressure receiving element. The piezoelectric element shown in the figure is radially polarized. The type of piezoelectric element used in this kind of sound receiver is shown separately in Fig. 101, but in this case the element is longitudinally polarized. The diaphragm 1 is also ceramic and covers one of the end surfaces of the cylinder; the silver electrodes 3 and 4 we deposited on the external curved surface of the ceramic cylinder 2 in the form of a double-start spiral.

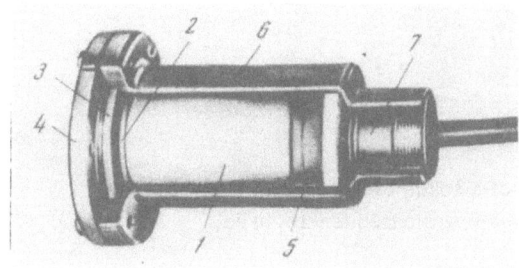

Fig. 100. Resonance sound receiver with plane receiver diaphragm and cylindrical element of barium titanate ceramic. 1) Cylindrical piezoelectric element; 2) end diaphragm; 3) rubber sealing ring; 4) clamping ring for seal; 5) rear rubber retainer; 6) outer casing; 7) cable gland.

Fig. 101. Longitudinally polarized ceramic piezoelectric element for a resonance sound receiver. 1) Diaphragm; 2) ceramic cylinder; 3, 4) electrodes.

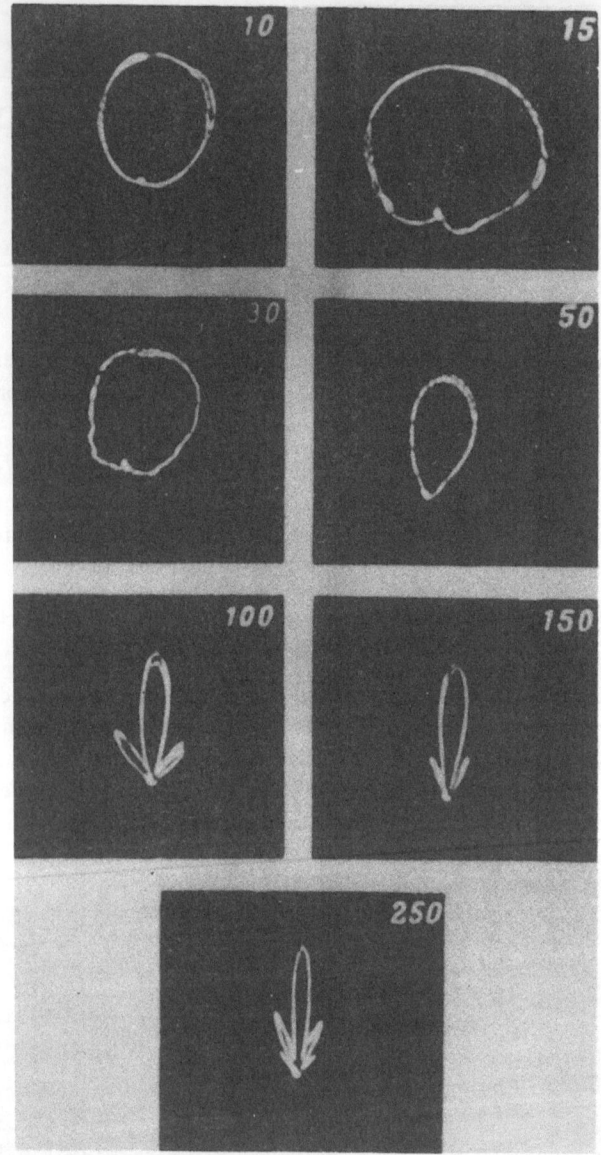

Fig. 102. Directivity characteristics of a sound receiver with a
plane receiver diaphragm. The figures give the frequencies in kc.

TABLE 23

No.	Type of polarization	f_{res}, kc	h, mm	C, $\mu\mu f \cdot 10^3$	l, mm	δ, mm	Resonance sensitivity, μV/bar
1	Longitudinal	26.8	60	5.7	6	3	47.6
2		38	38	3.1	6	1.5	40
3		24	60	2.82	12	3	84
4	Radial	20	60	53	1	1	7.4
5		20	60	56.5	1	1	6.7
6	Longitudinal	32	52	3.48	6	1.5	30
7		60	26	2	6	1.5	30
8		23.5	65	5.7	6	1.5	31
9		9.4	165	11.7	6	1.5	40
10	Radial	20	60	56.2	1	1	7.8
11		23	60	17	3	3	15.7
12	Longitudinal	12.5	125	10.3	6	1.5	19.4
13		14.5	85	6.8	6	1.5	46
14	Radial	20	60	57	1	1	7.7
15		24	60	–	5	5	40

Notation: f_{res} = resonance frequency; h = height of cylindrical piezoelectric element; l = distance between electrodes; δ = wall thickness of cylindrical element; C = capacity.

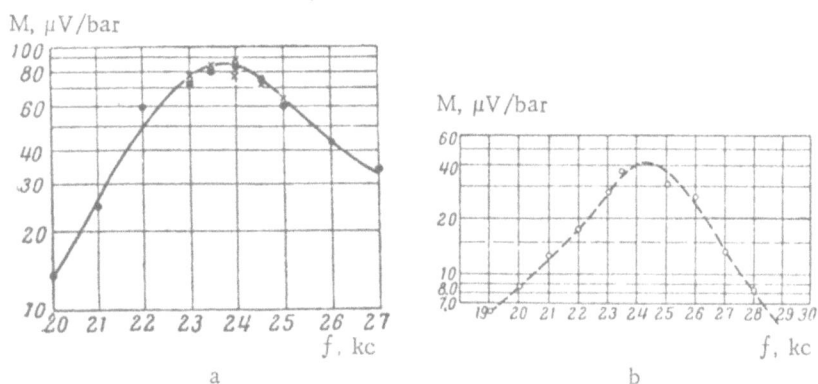

Fig. 103. Frequency characteristic of a resonance sound receiver with a plane receiver diaphragm with (a) longitudinally and (b) radially polarized cylindrical piezoelectric elements.

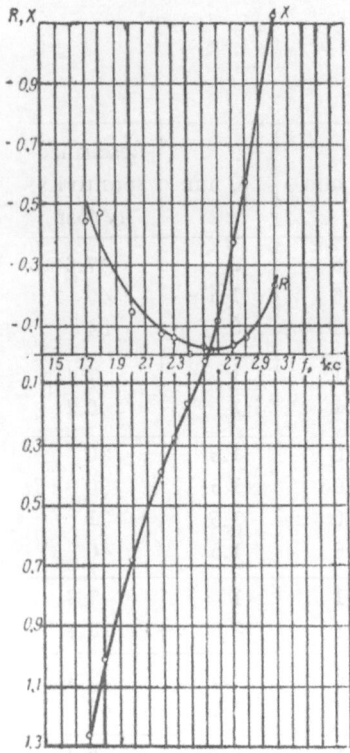

Fig. 104. Frequency characteristic of the total input acoustic impedance of a resonance sound receiver with a plane receiver diaphragm.

The parameters of the investigated transducers of this type are given in Table 23.

In all cases, the external diameter of the piezoelectric element was 3.2 cm. The sensitivity values given in the table were measured in a water-filled chamber (see Chapter II, Section 4).

Figure 102 shows the normal directivity characteristics for such a sound receiver with a resonance frequency of 25 kc. The form of these characteristics depends essentially on the frequency and the external diameter of the diaphragm and of course on the diffraction conditions.

The frequency characteristics of the sensitivity of such receivers were easily obtainable owing to the presence of the plane diaphragm. These characteristics were recorded either with the aid of an underwater sound duct or by means of a metal rod waveguide (see Chapter 2, Section 4).

By way of example, the frequency dependence of the sensitivity of a sound receiver with a resonance frequency of 24 kc and a longitudinally polarized piezoelectric element (Table 23, No. 3) is shown in Fig. 103a. This curve was obtained in a water-filled test chamber. Figure 103b gives the corresponding curve for a radially polarized sound receiver of the same resonance frequency (Table 23, No. 15).

The difference in resonance sensitivities measured in a water-filled chamber and by means of a solid waveguide amounted to approximately 2 db.

With the aim of using this type of transducer for different methods of calibration, it was of interest to determine the frequency behavior of the active and reactive components of the input acoustic impedance of the transducer. The corresponding measurements were carried out in a pulse tube [188]. The frequency dependence of the active (R) and reactive (X) components of the total input acoustic impedance of transducer No. 15 (Table 23) is shown in Fig. 104. It is clearly shown that the sound receiver becomes acoustically compliant at resonance. This fact must be taken into account, for example, when using such resonance receivers as receiver transmitters in calibration measurements using the reciprocal method, etc.

Sound receivers of the type described above possess relatively low directional properties due to the small diameter of the sound receiver diaphragm. Simply increasing the diaphragm diameter with the same piezoelectric element naturally results in a construction of the principal lobe of the directivity characteristic but may also result in enlargement of the secondary lobes. In the case of large diameters it would seem better to use a "mosaic" consisting of a number of cylindrical elements as indicated in Fig. 105, which shows a plane

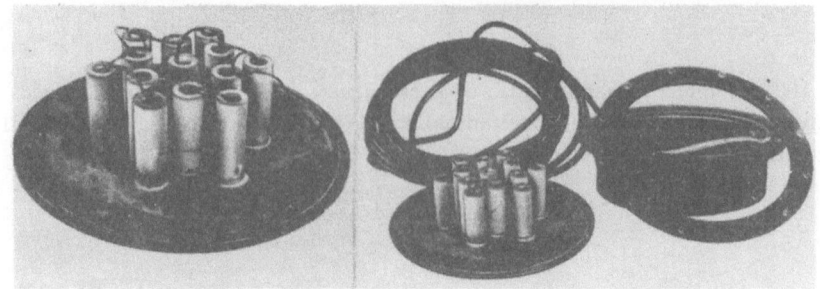

Fig. 105. "Mosaic" of ceramic cylinders with a resonance frequency of 27 kc.

108

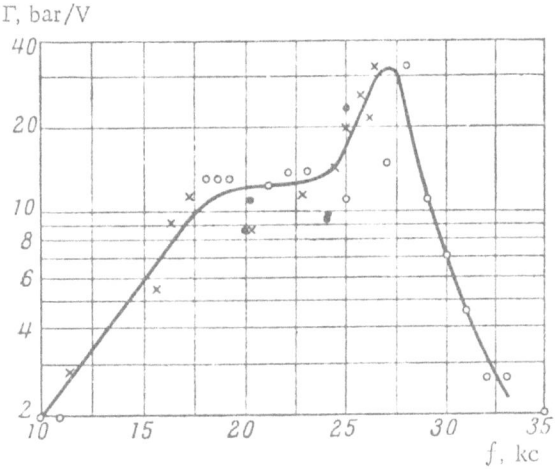

Fig. 106. Frequency response of a "mosaic" transducer
employing cylindrical elements.

steel diaphragm and 13 radially polarized ceramic cylinders; the cylindrical piezoelectric elements are carefully adjusted to the resonance frequency and secured to the diaphragm by their end faces with a carbinol adhesive. The diaphragm is mounted in the sound receiver housing by means of a rubber sealing ring.

Figure 106 shows the frequency characteristic for such a transducer with a diaphragm diameter of 32 cm and a resonance frequency of 27 kc acting as a transmitter.

We note that in this type of construction the rear end faces of the cylindrical elements remain free.

"Mosaic" transducers are of simple construction and can be produced to give practically any resonance frequency over the range from approximately 10 kc to 200 kc.

2. Resonance Sound Receivers with a Plane Piezoelectric Diaphragm

In many cases the very simple form of construction shown in Fig. 107 can be used. The thickness polarized plane diaphragm 1 of barium titanate ceramic is clamped along its edge to the metal housing 4 by means of the rubber ring 2 and the metal clamping ring 3. The external metallic coating simultaneously acts as a screen which is electrically connected to the housing. In most cases a protective film of vitreous enamel was deposited on top of the metallic layer. The lead from the internal electrode is brought out through a cable gland. This type of sound receiver was used for underwater sound purposes. The principal parameters of a number of such sound receivers are given in Table 24.

The sensitivity of these sound receivers was measured in a tank, the receivers being calibrated by the reciprocity method using sound pulses reflected from the free water surface.

It should be noted that the frequency characteristics of sound receivers with piezoelectric diaphragms vibrating in the direction of the diaphragm thickness frequently display irregularity in the splitting of the resonances, etc. This occurred in the present instance.

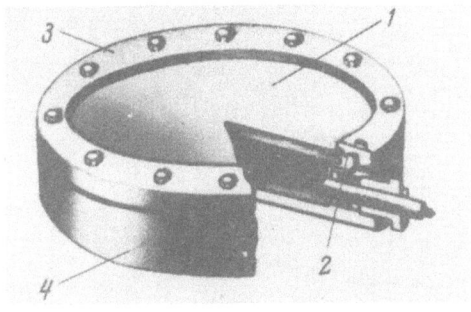

Fig. 107. Resonance sound receiver with a plane
piezoelectric half-wave receiver diaphragm.

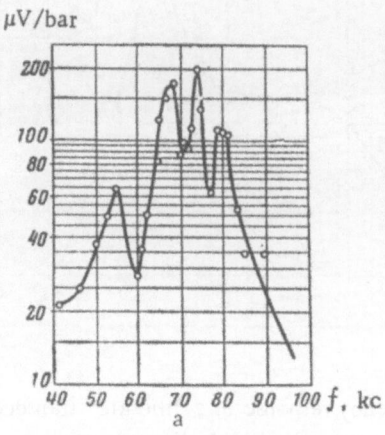

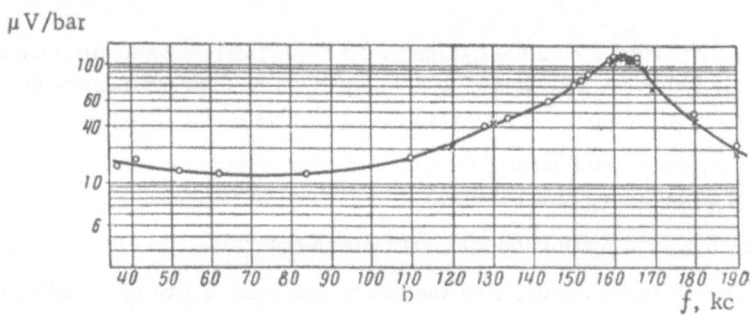

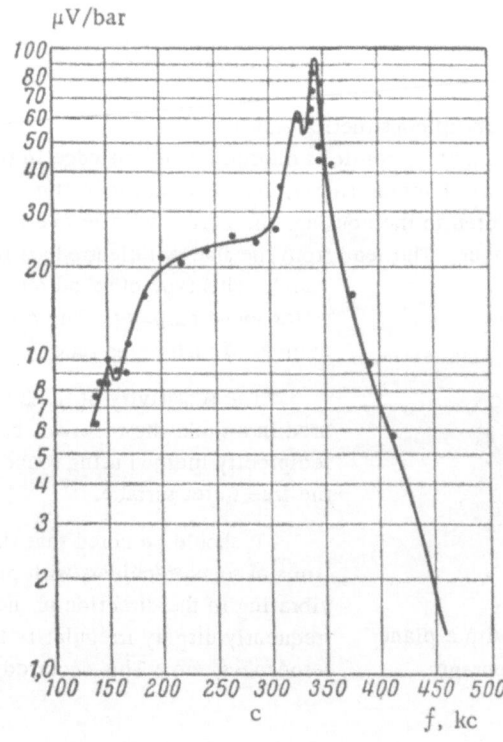

Fig. 108. Frequency characteristic of a sound receiver with a plane half-wave diaphragm of barium titanate ceramic. a) Diaphragm thickness 3.2 cm; b) 1.35 cm; c) 0.75 cm.

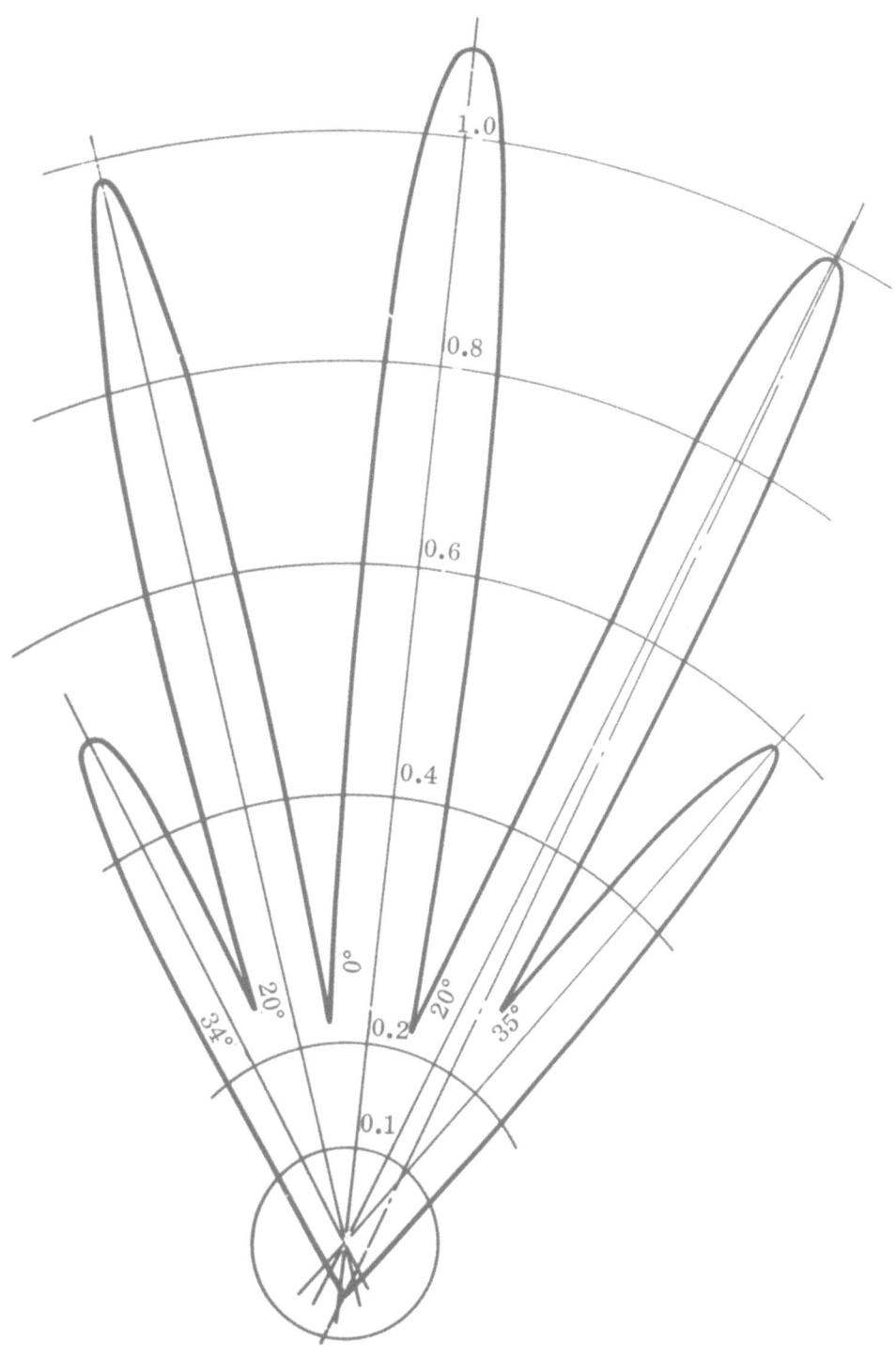

Fig. 109. Directivity characteristic of a transducer with a diaphragm in the form of a plane ceramic plate in which the vibrations of the surface are not cophasal.

TABLE 24

No.	Diaphragm dimensions		Resonance frequency, kc	Resonance sensitivity, μV/bar	Capacity, μμf
	Thickness, cm	Diameter, cm			
1	3.2	10.7	65- 75	170-200	3,500
2	1.35	12.2	160-165	120	7,800
3	0.75	12.7	349	95	16,000

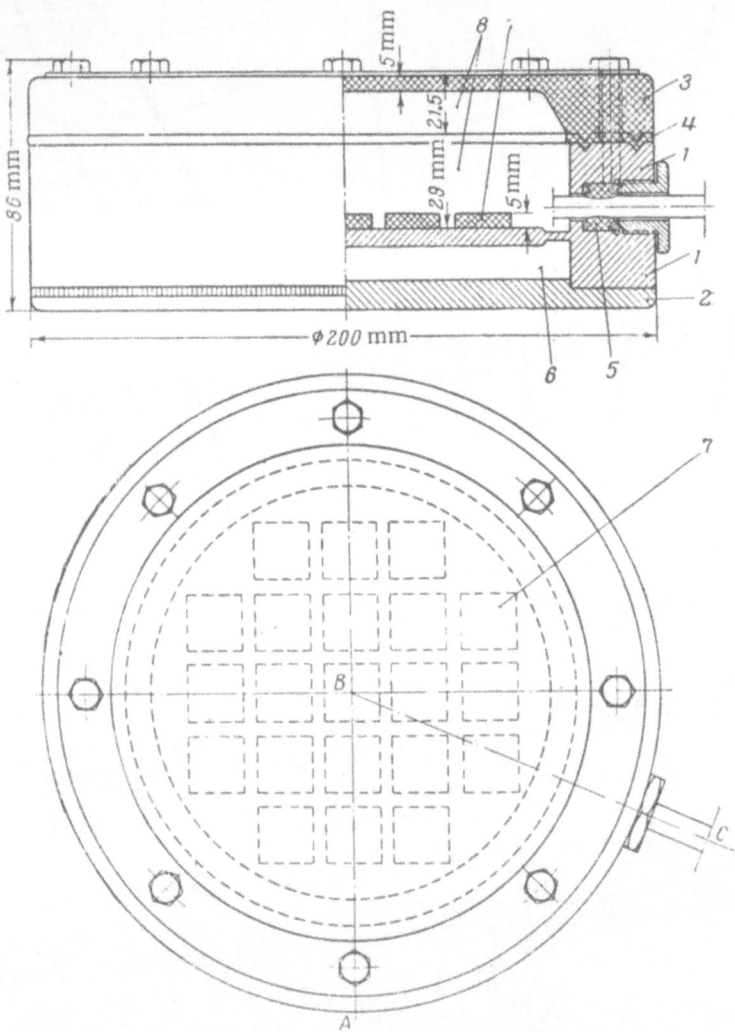

Fig. 110. Multilayer "mosaic"-type transducer with barium titanate ceramic elements.

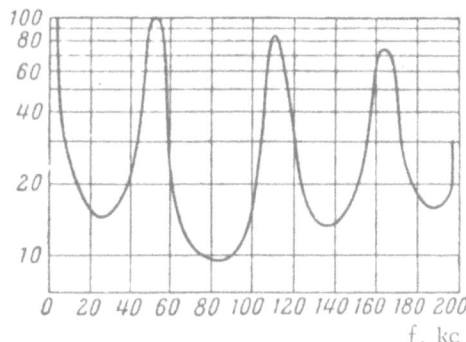

Fig. 111. Theoretical transmittance of a two-layer system equivalent to a one-sided transducer consisting of a ceramic element and a backing plate.

Figure 108a shows the frequency characteristic of sound receiver No. 1 (Table 24) with the thickest diaphragm (3.2 cm). This is a clear example of a complex spectrum in the region of the principal resonance. In the case of a moderate diaphragm thickness of 1.35 cm (Fig. 108b, sound receiver No. 2), the frequency characteristic possesses a very regular form while with a diaphragm thickness of 0.75 cm (Fig. 108c, sound receiver No. 3) irregularities once more appear in the form of the frequency characteristic.

These effects are presumably related to the complex nature of the vibration of a plane piezoelectric plate the thickness of which is small in comparison with the plate diameter and correspondingly to the complex distribution of the vibration amplitudes over the external surface of the plate [207, 208]. If the structure of the frequency characteristic is complex in the region of the principal resonance, it is difficult to speak of a frequency constant, Q-value, or width of the resonance curve. It may be that average values can be taken in such cases. Nevertheless, the distinguishing feature of ceramic sound receivers with diaphragms consisting of plane piezoelectric plates is their comparatively high Q-value. Therefore barium titanate may be considered a promising material for the construction of narrow band resonance transducers of the type under discussion. The high Q-value is due to the large coefficient of electromechanical coupling and high characteristic impedance of barium titanate ceramic.

The aperture of the principal lobe of the directivity characteristic of such transducers at various frequencies generally agrees with the theoretical value based on cophasal motion of the external surface of the diaphragm. However, at certain frequencies the secondary maxima of the directivity characteristics become much more pronounced in comparison with the values predicted by calculation. This is presumably caused by the existence of a system of transverse standing waves in the finite ceramic plate.

Figure 109 shows this type of marked distortion of the directivity characteristic for a sound receiver with a plane piezoelectric diaphragm 10 cm in diameter and 3.2 cm in thickness at a frequency of 150 kc. Transducers with plane piezoelectric diaphragms can be designed for use at relatively high frequencies.

In the case of transducers designed for use at lower frequencies it is more convenient to use an assembly-type of construction in which the plane piezoelectric element is loaded on one or both sides by metal backing plates.

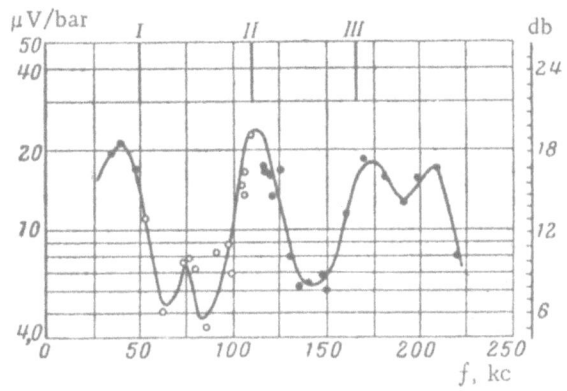

Fig. 112. Frequency dependence of a transducer consisting of a sensitive element and a backing plate.

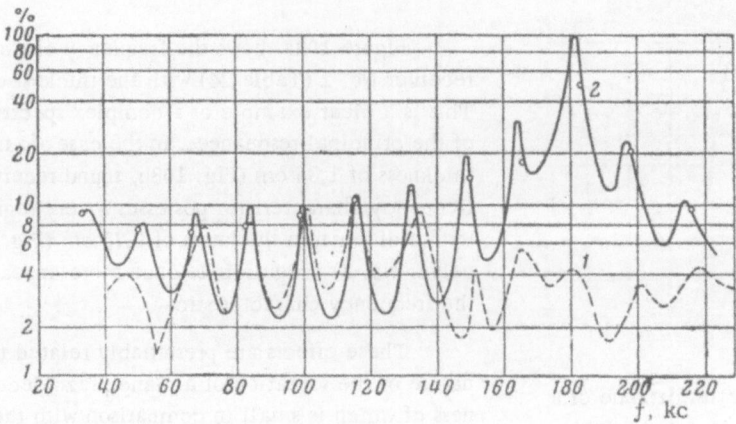

Fig. 113. Comparison of the frequency dependence of a multilayer transducer with the theoretical transmittance of an equivalent layer system.

If a transducer with a large diameter is required, the method of construction shown in Fig. 110 can be adopted. Here a mosaic of piezoelectric plates 7 with dimensions of $20 \times 20 \times 5$ mm^3 are attached by a carbinol adhesive to a circular steel plate 6 mm in thickness forming part of the cylindrical housing 1. All the piezoelectric elements are connected in parallel. The front of the housing is closed by a sound transmitting cover of perspex 3 and a rubber sealing ring 4. The internal cavity 8 is filled with transformer oil. The rear cover plate 2 is welded to the housing 1 so that an air-filled gap is formed between the rear cover plate and the plate carrying the ceramic elements. The cable is brought out through the gland 5.

These multilayer-type transducers are characterized by a complex natural frequency spectrum containing a number of resonances. The location of these resonances is easily determined by calculating the frequency characteristic of the acoustic transmittance of the layer assembly. Such calculations were carried out by us in accordance with the method proposed by Tartakovskii [210].

We compared the theoretical and experimental data [211] for a hydrophone the front surface of which consisted of a piezoceramic plate 1.3 cm in thickness which was backed by an aluminum disc 3.2 cm in thickness. The diameter of the assembly was 9.5 cm.

The theoretical frequency dependence of the acoustic transmittance of such a layer system (water−barium titanate−aluminum−air) is shown in Fig. 111 in which the transmittance is plotted along the ordinate axis in arbitrary units.

Figure 112 shows the frequency dependence of the sensitivity of such a sound receiver. The vertical lines indicate the locations of the first, second, and third theoretical transmittance maxima. The experimental natural frequency spectrum is in good agreement with the theoretical spectrum.

Figure 113 compares the experimental frequency dependence of the sensitivity of a "mosaic"-type transducer (see Fig. 110), (curve 1) with the theoretical frequency dependence of the transmittance of a perspex−oil−barium titanate−steel−air layer system (curve 2). In each case the ordinate scale is expressed in relative units. This figure clearly shows that even in the case of very complex layer systems good agreement is obtained between the theoretical and experimental natural frequency spectra [212, 213].

Thus, the use of a backing plate enables the construction of transducers with wide natural frequency spectra including much lower natural frequencies than could be obtained by the use of the piezoelectric plate alone. However, the resonance sensitivity of a transducer using a backing plate is always somewhat lower than the sensitivity of a pure ceramic transducer (by 1.5 to 2 times) presumably due to losses in the adhesive or other intermediate layers.

114

LITERATURE CITED

1. M. P. Langevin, French Patents Nos. 505703/1918, 576281/1924, 622035/1926; Sounding with the aid of sound, Hydrogr. Rev. 2 (2): 51-121 (1924); Ultrasonic sounding apparatus, Special Publ. Int. Hydrogr. Bureau, No. 14, 1926.

2. A. M. Nicolson, Proc. Am. Inst. Elec. Engrs. 38 (11): 1315-1333 (1919); Piezoelectric effect in a complex Rochelle salt crystal, Trans. Am. Inst. Elec. Engrs. 38: 1467 (1919).

3. E. Klein, Early stages in the development of ultrasonics, J. Acoust. Soc. Am. 20 (5): 601 (1948).

4. C. B. Sawyer, Use of Rochelle salt crystals in reproducers and microphones, Proc. Inst. Radio Engrs. 19: 2020-2029 (1931).

5. N. N. Andreev, Inventor's Certificate No. 27406 1930.

6. A. L. Williams, Electronics 4: 166-167 (1932); 25: 196 (1934); Piezoelectric loudspeakers and microphones, J. Soc. Motion Picture Engrs. 23: 1420-1421 (1935).

7. S. Ballantine, Piezoelectric loudspeakers for use at high frequencies, Proc. Inst. Radio Engrs. 21: 1399-1408 (1933).

8. S. Ballantine, High quality broadcasting transmission and reception, Proc. Inst. Radio Engrs. 22: 564-629 (1934); 23: 1420-1421 (1935).

9. A. L. Williams and I. P. Arndt, Design of directional crystal microphones, Electronics 8: 242-243 (1935).

10. T. Sugimoto, A Rochelle salt crystal microphone and an underwater acoustic detector, Rep. Radio Research, Japan 6: 9-10 (1936).

11. P. Beerwald and H. Keller, Piezoelectric crystal elements for electroacoustic purposes (including Rochelle salt bimorphs), Funktechn. Monatshefte, 11: 345-348 (1938).

12. P. Beerwald and H. Keller, Funktechn. Monatshefte 12: 187-190 (1941); Theory and practice of piezoelectric microphones, Wireless Eng. 19: 1393A (1942).

13. H. Keller, Piezoelectric "vibrating strip" as an electromechanical transducer, Hochfrequenztechn. u. Electroac. 60: 5-10 (1942).

14. R. S. Sawdeg, Bimorph Rochelle salt piezoelectric elements and their application, Radio, Vol. 23 et seq. (Sept. 1943).

15. A. A. Kharkevich, The use of Rochelle salt in piezoelectric devices, Zh. Tekhn. Fiz. 13: 585 (1943).

16. V. N. Lepeshinskaya, Piezoelectric Devices Using Rochelle Salt, Leningrad, 1943.

17. W. P. Mason, Elastic, piezoelectric, and dielectric constants of potassium dihydrogen phosphate and ammonium dihydrogen phosphate, Phys. Rev. 69: 173 (1946); Properties of monoclinic crystals, Phys. Rev. 70: 9-10 (1946).

18. R. Bechman, Determination of the elastic and piezoelectric coefficients of monoclinic crystals including ethylene−diamine−tartrate, Proc. Phys. Soc. 63b: 577 (1950).

19. F. Spitzer, Piezoelectric crystals, Fernmeldetechn. Z. 3: 190-196 (1950); 3: 234-243 (1950).

20. F. Spitzer, New synthetic piezoelectric crystals in electroacoustics and high frequency techniques, Arch. Elektr. Übertrag. 5: 544-554 (1951).

21. W. Cady, Piezoelectricity and Its Practical Application, IL, 1949.

22. W. P. Mason, Piezoelectric Crystals and Their Application in Ultrasonics, IL, 1952.

23. R. Bechman, Elastic and piezoelectric coefficients of lithium sulfate monohydrate, Proc. Phys. Soc. 65 (5): 389 (1952).

24. N. N. Andreev, Piezoelectric crystals and their application, Elektrichestvo 2: 5-13 (1947).

25. L. Fein, An ultrasonic underwater "point source" probe, J. Acoust. Soc. Am. 20: 590 (1948); 22: 876 (1950).

26. L. Rudnick and H. C. Rothenberg, A miniature high frequency microphone, J. Acoust. Soc. Am. 20: 594 (1948).

27. W. Günter, A crystal microphone for underwater acoustic measurement, Z. Angew. Phys. 2: 206 (1950).

28. H. W. Cooper, A miniature crystal transducer as a probe for ultrasonic investigations, J. Acoust. Soc. Am. 22: 86 (1950).

29. P. V. Anan'ev, Synthetic crystals and their application to instrumentation, Tr. Komis. po Akustike 6: 3 (1951).

30. B. M. Vul and I. M. Gol'dman, The dielectric constants of titanates of second group metals, Dokl. Akad. Nauk SSSR 46 (4): 154 (1945).

31. B. M. Vul and I. M. Gol'dman, The dielectric constant of barium titanate as a function of the alternating field intensity, 49 (3): 179-182 (1945).

32. B. M. Vul and I. M. Gol'dman, Dielectric hysteresis in barium titanate, Dokl. Akad. Nauk SSSR 51 (1): 21 (1946).

33. B. M. Vul and L. F. Vereshchagin, Dependence of the dielectric constant of barium titanate on pressure, Dokl. Akad. Nauk SSSR 48 (9): 662 (1945).

34. B. M. Vul, Dielectric constant of barium titanate at low temperatures, Zh. Eksperim. i Teor. Fiz. 15 (12): 735-736 (1945).

35. B. M. Vul, Materials with high and extra high dielectric constants, Elektrichestvo, No. 3 (1946).

36. J. Bretteville, Oscillograph investigations of the dielectric properties of barium titanate, J. Am. Ceramic Soc. 29: 303 (1946).

37. R. Adler and W. L. Cherry, Piezoelectric effect in polycrystalline barium titanate, Phys. Rev. 72 (10): 981 (1947).

38. S. Roberts, Dielectric and piezoelectric properties of barium titanate, Phys. Rev. 71: 890 (1947).

39. B. M. Vul and I. M. Gol'dman, A new type of barium titanate, Dokl. Akad. Nsuk SSSR 60 (1): 41 (1948).

40. A. V. Rzhanov, Spontaneous Polarization in the Piezoelectric Effect in Barium Titanate, Thesis, FIAN, 1948.

41. A. V. Rzhanov, Spontaneous polarization in $BaTiO_3$ specimens, Zh. Éksperim. i Teor. Fiz. 19 (4): 335 (1949).

42. A. V. Rzhanov, Piezoelectric effect in barium titanate, Zh. Éksperim. i Teor. Fiz. 19 (6): 502-506 (1949).

43. A. V. Rzhanov, Barium titanate, a new ferroelectric, Usp. Fiz. Nauk 38 (4): 486-487 (1949).

44. A. A. Anan'eva and V. M. Tsarev, Development of nondirectional sound receivers for use at ultrasonic frequencies, Report of the Acoustic Laboratory, FIAN, 1951.

45. L. G. Hector and H. W. Koren, Ceramic sound receivers, Electronics, pp. 94-96 (1948).

46. G. N. Howatt, J. W. Crowner, and A. Drametz, New Piezoelectric Materials, Electronics pp. 97-99 (1948).

47. J. Bugosh, E. Yeager, and F. Hovork, A high frequency barium titanate hydrophone, Phys. Rev. 76 (2): 1890 (1950).

48. E. Ackerman and W. Holak, A ceramic measurement microphone, Rev. Sic. Instr. 25: 857-861 (1954).

49. H. W. Koren, Application of polarized ceramics in transducers, J. Acoust. Soc. Am. 21 (3): 198-201 (1949).

50. W. P. Mason, Barium titanate ceramic as an electromechanical transducer, Bell Lab. Rec. 27: 285-289 (1949).

51. W. P. Mason and R. F. Wick, Barium titanate ceramic transducer for large displacements at ultrasonic frequencies, J. Acoust. Soc. Am. 23: 209-214 (1951).

52. T. F. Johnston and F. D. Wertz, A cylindrical barium titanate transducer, J. Acoust. Soc. Am. 22: 676 (1950).

53. G. Bradfield, Ultrasound in solids, Res. Sci. Appl. J. 6 (2): 68-79 (1953).

54. L. Camp, Wideband barium titanate transmitter, J. Acoust. Soc. Am. 25: 297-301 (1953).

55. O. Matiat, Transducers for generating ultrasonic waves, J. Acoust. Soc. Am. 25: 991-996 (1953).

56. T. F. Hueter, A temperature invariant impedance point in the frequency response of barium titanate transducers, J. Acoust. Soc. Am. 25: 152 (1953).

57. H. Jaffe, Titanate ceramics for electromechanical applications, Ind. Eng. Chem. 42 (2): 264-268 (1950).

58. R. S. Roth, Elementary cell data for lead niobate, Acta Cryst. 10: 437 (1951).

59. G. Shirane and K. Suzuki, Crystal structure, J. Phys. Soc. Japan 7 (3): 333 (1952).

60. E. Sewaguchi, Ferroelectric and antiferroelectric properties of a $PbZrO_3 - PbTiO_3$ solid solution, J. Phys. Soc. Japan 5 (8): 615-629 (1953).

61. G. Goodman, Ferroelectric properties of lead metaniobate, J. Am. Ceram. Soc. 36: 368 (1953).

62. B. Jaffe, R. S. Roth, and S. Mazzullo, Piezoelectric properties of ceramics based on solid solutions of lead zirconate and lead titanate, J. Appl. Phys. 6: 25 (1954).

63. W. P. Mason, The application of barium titanate ceramics stable in regard to aging and temperature in transducers, in the excitation of elastic waves and for the measurement of forces, Acustica 1: 200 (1954).

64. B. Jaffe, R. S. Roth, and S. Mazzullo, Properties of piezoelectric ceramics in the form of solid solutions, J. Research Natl. Bur. Standards 55 (5): 239 (1955).

65. L. Egertone and S. E. Koonce, Effect of annealing on the structure and dielectric and piezoelectric properties of barium titanate ceramic, J. Am. Ceram. Soc. 38 (11): 413 (1955).

66. D. Berlincourt, Recent developments in the field of ferroelectric materials for transducers, IRE Trans. Ultrasonic Eng. 4: 53-65 (1956).

67. R. Bechman, Elastic, piezoelectric, and dielectric constants of polarized barium titanate ceramic and some applications of the piezoelectric equations, J. Acoust. Soc. Am. 28 (3): 347-350 (1956).

68. M. H. Francombe and B. Lawis, Structure and phase transitions in ferroelectric sodium−lead niobates and other sodium−niobate ceramics, Electronics 2 (4): 387-403 (1957).

69. D. Berlincourt and H. U. Krueger, Dependence of the ratio of the piezoelectric constants on density and composition in barium titanate ceramics, Phys. Rev. 105 (1): 56-57 (1957).

70. J. M. Herbert, Ferroelectric crystals and ceramics, J. Electron. Control, Ser. I, 1 (5): 2 (1958).

71. A. E. Crawford, Piezoelectric lead zirconate ceramics, Brit. Commun. Electron. 6 (7): 516-519 (1959).

72. F. Kulesar, Electromechanical properties of a lead zirconate titanate ceramic in which the lead is partially substituted by calcium and strontium, J. Am. Ceram. Soc. 42 (1): 49 (1959).

73. F. Kulesar, Electromechanical properties of a lead zirconate titanate ceramic modified by the addition of trivalent and tetravalent materials, J. Am. Ceram. Soc. 42 (7): 343 (1959).

74. L. Egerton and D. M. Dillon, Piezoelectric and dielectric properties of sodium niobate−potassium niobate ceramics, J. Am. Ceram. Soc. 42 (9): 438 (1959).

75. G. A. Smolenskii, Ferroelectric properties of certain titanates and zirconates of divalent metals possessing perovskite structure, Zh. Tekhn. Fiz. 20 (2): 137 (1950).

76. G. A. Smolenskii, New ferroelectrics, Dokl. Akad. Nauk SSSR 70 (3): 405-407 (1950).

77. G. A. Smolenskii, Electrostriction effects in ceramic ferroelectrics, Zh. Tekhn. Fiz. 21 (9): 1045-1049 (1951).

78. G. A. Smolenskii and I. V. Kozhevnikova, The formation of ferroelectrics, Dokl. Akad. Nauk SSSR 76: 519-522 (1951).

79. G. A. Smolenskii, M. A. Karamyshev, and K. I. Rozgachev, Ferroelectric properties of certain solid solutions, Dokl. Akad. Nauk SSSR 79: 53 (1951).

80. G. A. Smolenskii, Ferroelectric properties of certain crystals, Dokl. Akad. Nauk SSSR 85 (5): 985-987 (1952).

81. G. A. Smolenskii, Ferroelectrics with Perovskite Structure, Thesis, Leningrad, 1954.

82. G. A. Smolenskii and V. A. Isupov, Ferroelectric properties of barium stannate−barium titanate solid solutions, Zh. Tekhn. Fiz. 24 (8): 1373-1376 (1954).

83. G. A. Smolenskii, N. T. Tarutin, and N. P. Grudtsin, Ferroelectric properties of barium zirconate−barium titanate solid solutions, Zh. Tekhn. Fiz. 24: 1584 (1954).

84. G. A. Smolenskii and K. I. Roznachev, Ferroelectric properties of barium titanate−strontium titanate solid solutions, Zh. Tekhn. Fiz. 24 (10): 1753 (1954).

85. G. A. Smolenskii, A. N. Agranovskaya, A. M. Kalinin, and T. M. Fedotova, Ferroelectric properties of solid solutions, Zh. Tekhn. Fiz. 25 (12): 2134-2142 (1955).

86. G. A. Smolenskii, New ferroelectrics and antiferroelectrics of the oxygen−octahedral type, Izv. Akad. Nauk SSSR, Ser. Fiz. 20 (2): 163 (1956).

87. G. A. Smolenskii, V. A. Isupov, and A. N. Agranovskaya, Phase transitions in the ferroelectric state of strontium pyrotantalate, Dokl. Akad. Nauk SSSR 113 (4): 803-804, 1053-1056 (1957).

88. G. A. Smolenskii, New ferroelectrics and antiferroelectrics, Usp. Fiz. Nauk 62 (1): 41-49 (1957).

89. N. A. Roi, Electromechanical Properties of Barium Titanate and Certain Solid Solutions Based on Barium Titanate, Thesis, Moscow, AKIN, 1953.

90. N. A. Roi, Dielectric and piezoelectric properties of the solid solutions $(BaSr)TiO_3$, $Ba(TiSn)O_3$, and $(TiZr)O_3$, Akust. Zh. 2 (1): 62-70 (1956).

91. V. A. Bokov, Piezoelectric properties of the polycrystalline solid solutions $(BaSn)TiO_3$, $Ba(TiSn)O_3$, and $Ba(TiZr)O_3$, Akust. Zh. 3 (2): 104-108 (1957).

92. V. A. Bokov, The behavior of ferroelectrics in strong electric fields, Izv. Akad. Nauk SSSR, Ser. Fiz. 21 (3): 382 (1957).

93. I. E. Myl'nikova, Investigation of solid solutions possessing ferroelectric properties, Izv. Akad. Nauk SSSR, Ser. Fiz. 21 (3): 423-432 (1957).

94. V. A. Isupov and V. I. Kosyakov, Dielectric polarization and piezoelectric properties of ferroelectric solid solutions of metaniobates of calcium, strontium, and barium with lead metaniobate, Zh. Tekhn. Fiz. 28 (10): 2175 (1958).

95. V. A. Isupov and V. I. Kosyakov, Dielectric polarization and piezoelectric properties of certain ferroelectric solid solutions based on sodium niobate, Fiz. Tverd. Tela. 1 (6): 929 (1959).

96. R. E. Pasynkov and V. V. Vinogradov, Stabilized piezoelectric materials, Izv. Akad. Nauk SSSR, Ser. Fiz. 21 (3): 450-454 (1957).

97. H. D. Megaw, Crystalline structure of binary oxides of the perovskite type, Proc. Phys. Soc. 258 (326): 133 (1946).

98. H. D. Megaw, Temperature variations in the crystalline structure of barium titanate, Proc. Roy. Soc. A189: 261 (1947).

99. H. D. Megaw, Crystalline structure of barium titanate, Nature 155 (3938): 484-485 (1945).

100. H. D. Megaw, Changes in polycrystalline strontium titanate at the phase transition temperatures, Nature 157: 20-21 (1946).

101. B. Matthias and A. Hippel, Domain structure and dielectric properties of barium titanate single crystals, Phys. Rev. 73: 1378 (1948).

102. G. Danielson, B. Matthias, J. Richardson, and I. Bernd, Properties of single domain $BaTiO_3$ crystals, Phys. Rev. 74 (8): 986-987 (1948).

103. W. Merz, Dielectric properties of $BaTiO_3$ crystals, Phys. Rev. 75 (4): 687 (1949).

104. W. Merz, Effect of hydrostatic pressure on the Curie point of barium titanate crystals, Phys. Rev. 78 (1): 52-54 (1950).

105. G. Schirane and A. Takeda, Variations in volume accompanying the three phase transitions in $BaTiO_3$, J. Phys. Soc. Japan 6: 128-129 (1950).

106. W. Könzig and N. Maikoff, Is the ferroelectric phase transition at 120°C a transition of the first or second kind?, Helv. Phys. Acta 24 (4): 342-356 (1951).

107. M. E. Drougard and D. R. Joung, Effect of domain orientation in barium titanate single crystals, Phys. Rev. 94 (15): 1561-1564 (1954).

108. M. E. Drougard and D. R. Joung, Dielectric properties of single domain barium titanate crystals in the neighborhood of the Curie point, Phys. Rev. 95 (5): 1152-1153 (1954).

109. N. S. Novopel'tsev and A. A. Khodakov, Dielectric hysteresis in $BaTiO_3$ single crystals, Zh. Eksperim. i Teor. Fiz. 27 (1): 94-96 (1954).

110. P. W. Forsberg, Effect of two-way pressure on the Curie point in barium titanate, Phys. Rev. 93 (4): 686-692 (1954).

111. M. E. Drougard, H. L. Funk, and D. R. Joung, Dielectric constant and losses measured on the basis of hysteresis loops for crystalline barium titanate specimens, J. Appl. Phys. 25 (9): 1166-1169 (1954).

112. K. W. Plessuer and K. A. Cook, Anomalies of the temperature coefficient of permeability in barium titanate, Nature 4406: 682-683 (1954).

113. H. H. Wieder, Ferroelectric hysteresis in barium titanate crystals, J. Appl. Phys. 26 (12): 1479-1482 (1955).

114. A. Mizarova, Temperature hysteresis of the dielectric constant of a single crystal, Zh. Tekhn. Fiz. 6 (5): 527-529 (1956).

115. W. L. Bond, W. Mason, and H. McShimin, Elastic constants and electromechanical coupling coefficients of barium titanate, Phys. Rev. 82 (3): 442 (1951).

116. B. M. Vul, I. M. Gol'dman, and R. Ya. Rozbash, The electrical strength of titanates of metals of the second group of the Mendeleev Table, Zh. Eksperim. i Teor. Fiz. 20 (5): 465-473 (1950).

117. B. M. Vul, S. V. Bogdanov, and R. Ya. Rozbash, Effect of polarization conditions on the piezoelectric properties of barium titanate, FIAN report, 1952.

118. B. M. Vul, S. V. Bogdanov, and R. Ya. Rozbash, Effect of polarization conditions on the piezoelectric properties of barium titanate, Zh. Tekhn. Fiz. 26: 5 (1956).

119. A. A. Karpacheva (Anan'eva), Application of the induced piezoelectric effect in barium titanate to acoustic purposes, FIAN report, 1949.

120. I. S. Zheludev, L. Z. Rusakov, L. A. Shuvalov, and M. M. Kachkacheva, Low temperature polarization of barium titanate ceramics, Izv. Akad. Nauk SSSR, Ser. Fiz. 22 (12): 1516 (1956).

121. B. M. Vul and S. V. Bogdanov, Effect of large mechanical loads on the piezoelectric properties of polarized barium titanate specimens, FIAN report, 1952.

122. B. M. Vul, S. V. Bogdanov, and R. Ya. Rozbash, Dependence of the transverse piezoelectric constant (d_{31}) of polarized barium titanate specimens on pressure, FIAN report, 1952.

123. E. K. Dovrer and K. N. Karmen, Characteristics of the piezoelectric effect in barium titanate under static conditions, Zh. Tekhn. Fiz. 27 (3): 513 (1957).

124. I. P. Kozlabaev, Piezoelectric properties of barium titanate, FIAN report, 1950.

125. A. V. Rzhanov, Piezoelectric and elastic properties of barium titanate under dynamic conditions, FIAN report, 1950.

126. N. A. Roi, The 120°C phase transition in barium titanate, Dokl. Akad. Nauk SSSR, Ser. Fiz. 81 (4): 545-547 (1951).

127. A. Piekara and Z. Pajak, Thermal pseudohysteresis of the dielectric constant in ferroelectric titanates, Acta Phys. Polon. 11 (3-4): 256-262 (1951-1952).

128. D. Ya. Kazarnovskii, Variation of the dielectric properties of ferroelectric ceramics with time, Zh. Tekhn. Fiz. 22: 553-558 (1952).

129. A. L. Khodakov, Dielectric losses in ceramic dielectrics and barium titanate at high frequencies, Zh. Tekhn. Fiz. 20 (5): 529-533 (1950).

130. A. Piekara and Z. Pajak, Acta Phys. Polon. 12 (3-4): 170-180 (1953).

131. F. S. Zavel'skii, Dependence of the dielectric constant of barium titanate on the duration of the effect of the voltage, Zh. Tekhn. Fiz. 25: 479 (1953).

132. K. W. Plessuer and K. A. Cook, Anomalies in the temperature coefficient of permeability in barium titanate, Nature 173: 682-683 (1954).

133. M. E. Drougard, R. Landuer, and D. R. Joung, Dielectric properties of barium titanate in the paraelectric state, Phys. Rev. 98 (4): 1010 (1955).

134. Jezewski and Piech, Dielectric properties of $BaTiO_3$ ceramic, Acta Phys. Polon. 14: 395 (1955).

135. G. Mesuard and L. J. Eyrand, Dielectric properties of barium titanate ceramic, J. Phys. Radium 17 (Suppl. 6): 62A, 78-80 (1956).

136. E. V. Sinyakov and V. V. Gal'perin, Dependence of the dielectric constant and tangent of the dielectric losses of barium titanate on the intensity of the high frequency electric field, Zh. Eksperim. i Teor. Fiz. 30: 675 (1956).

137. I. A. Itskhoki, Effect of pressure on the dielectric characteristics of ferroelectric ceramics, Izv. Akad. Nauk SSSR, Ser. Fiz. 20 (2): 199-200 (1956).

138. C. Feldman, Time changes in $BaTiO_3$, J. Appl. Phys. 27 (8): 870-873 (1956).

139. M. S. Lur'e and A. I. Medovoi, Effect of pressure on the dielectric constant, Zh. Eksperim. i Teor. Fiz. 26: 1437 (1956).

140. G. I. Skanavi and G. A. Lipaeva, Dielectric constant and loss angle of certain solid dielectrics at a wavelength of 3 cm and their dependence on temperature and frequency, Zh. Eksperim. i Teor. Fiz. 30: 824 (1956).

141. M. E. Drougard and E. J. Huibregtse, Effect of an electric field on the transitions in barium titanate, J.B.M.J. Res. Develop. 4: 318-329.

142. H. Robendorst and J. Melichercik, Dielectric properties of barium titanate at high frequencies and low temperature, Ann. Phys. Fol. 7 (1): 4-5, 261, 263 (1958).

143. Ya. M. Ksendzov and B. A. Rotenberg, Effect of pressure on the electrical properties of barium titanate in weak fields, Zh. Tekhn. Fiz. 1: 4 (1959).

144. J. Takagi, F. Sawaguchi, and T. Akioka, Effect of mechanical stresses on the permeability of barium titanate, J. Phys. Soc. Japan 3: 270 (1948).

145. J. Valasek, Piezoelectric activity of Rochelle salt under various conditions, Phys. Rev. 19: 478-491 (1922); Properties of Rochelle salt related to the piezoelectric effect, Phys. Rev. 20: 639-664 (1922).

146. I. P. Kozlabaev, Stability of the piezoelectric effect in polycrystalline barium titanate, Izv. Akad. Nauk SSSR, Ser. Fiz. 22 (2): 185 (1956).

147. W. P. Mason, Aging of the properties of barium titanate and related ferroelectric ceramics, J. Acoust. Soc. Am. 27: 1 (1955).

148. B. M. Vul, S. V. Bogdanov, and R. Ya. Rozbash, Piezoelectric shear coefficient (d_{15}) in polarized barium titanate, FIAN report, 1952.

149. E. G. Bronnikova and N. M. Stavitskii, Industrial piezoelectric materials and their application, "Piezotekhnika" (Tr. Tsentr. Nauchn.-Issled. Lab. P'ezotekhniki), Vol. 1, 1956.

150. S. V. Bogdanov, B. M. Vul, and A. M. Timonin, Relationship between the dielectric, piezoelectric, and elastic properties of a polycrystalline ceramic and a single crystal, Izd. Akad. Nauk SSSR, Ser. Fiz. 21 (3): 374 (1957).

151. S. V. Bogdanov, B. M. Vul, and R. Ya. Rozbash, Piezoelectric shear coefficient of polarized barium titanate, Kristallografiya 2 (1): 115-118 (1957).

152. G. A. Smolenskii and V. A. Zhukov, Ferroelectrics, Leningrad, Published by Obshchestvo po Rastprostraneniyu Politicheskikh i Nauchnykh Znanii, RSFSR, 1957.

153. D. Berlincourt and H. Jaffe, Elastic and piezoelectric coefficients of barium titanate, Phys. Rev. 3: 1 (1958).

154. I. S. Rez, E. G. Smazhevskaya, and M. M. Kachkacheva, The production of piezoceramics for high temperature use, Izd. Akad. Nauk SSSR, Ser. Fiz. 22 (12): 1520-1523 (1958).

155. I. L. Serova, V. S. Sluchevskii, and P. L. Strelets, Production of Ceramic Piezoelectric Elements, Sudpromgiz, 1959.

156. The Future of Ceramics in French Industry, Moscow, Exhibition, 1960.

157. A. A. Anan'eva, Sound Receivers of Barium Titanate Ceramic, Thesis, Kiev, 1961.

158. A. A. Anan'eva, M. A. Ugryumova, and B. V. Strizhkov, Report Akust. Inst. Akad. Nauk SSSR, 1959.

159. A. A. Anan'eva, M. A. Ugryumova, and B. V. Strizhkov, Dielectric and piezoelectric properties of chemically pure barium titanate ceramics, Izd. Akad. Nauk SSSR, Ser. Fiz. 24: 11 (1960).

160. V. P. Konstantinova, I. M. Sil'vestrova, and K. S. Aleksandrov, The production of triglycine sulfate crystals and their physical properties, Kristallografiya 4 (1): 69 (1959).

161. V. P. Konstantinova, I. M. Sil'vestrova, and V. A. Yurin, Twinning and dielectrical properties of triglycine sulfate crystals, Kristallografiya 4 (1): 125 (1959).

162. I. M. Sil'vestrova, K. S. Aleksandrov, and A. A. Chumakov, Growth of terpene monohydrate crystals and their elastic and piezoelectric properties, Kristallografiya 3 (3): 386 (1958).

163. A. A. Chumakov, I. M. Sil'vestrova, and K. S. Aleksandrov, Dielectric, elastic, and piezoelectric properties of benzophenone, Kristallografiya 2 (5): 707 (1957).

164. A. A. Chumakov, I. M. Sil'vestrova, and K. S. Aleksandrov, Growth of L-rhamnose monohydrate crystals and investigation of their dielectric properties, Kristallografiya 3: 4 (1958).

165. A. A. Anan'eva, Piezoceramic materials, FIAN report, 1951.

166. I. S. Zheludev, Thesis, Inst. Kristallografiya, Akad. Nauk SSSR, 1953.

167. I. S. Zheludev, I. M. Sil'vestrova, and V. P. Konstantinova, Investigation of Piezoelectric Orientations, Izd. Akad. Nauk SSSR, Moscow, 1955.

168. D. Berlincourt, B. Jaffe, H. Jaffe, and H. Krueger, Transducer properties of lead titanate—zirconate ceramic, IRE Natl. Convent. Rec. 6: 227-240 (1959).

169. H. Jaffe, Piezoceramics, J. Am. Ceram. Soc. 41 (11): 494-498 (1958).

170. O. I. Prokopalo and E. G. Fesenko, An investigation of the properties of solid solutions of the titanate and ferrite of barium and lead, Izd. Akad. Nauk SSSR, Ser. Fiz. 22: 12 (1958).

171. W. W. Malinofsky and H. Kedesdy, Barium and iron oxides isomorphic with hexagonal and tetragonal $BaTiO_3$, J. Am. Ceram. Soc. 76: 3090 (1954).

172. H. L. Yakel, Structure of certain compounds with perovskite type structures, Acta Cryst. 8: 394 (1955).

173. E. G. Fesenko and O. I. Prokopalo, Isomorphism of titanates and ferrates of strontium, barium, and lead, Kristallografiya 1 (5): 520 (1956).

174. D. Schofield and R. F. Brown, Investigation of a number of barium titanate compositions for use as transducers, Can. J. Phys. 35 (5): 594-607 (1957).

175. W. R. Cook and H. Jaffe, Ferroelectricity in oxides of fluorite structure, Phys. Rev. 88: 1426 (1952).

176. I. A. Trifonov, High temperature polymorphic transformations in barium titanate, Izv. Akad. Nauk SSSR, Ser. Fiz. 21: 12 (1958).

177. A. A. Anan'eva, M. A. Ugryumova, and B. V. Strizhkov, Production and anomalous properties of chemically pure barium titanate ceramic, paper read at fourth All-Soviet Conf. on Ferroelectrics, Jan. 1960.

178. B. M. Vul, Production and investigation of polycrystalline barium titanate piezoelectrics, FIAN report, 1950.

179. V. S. Belevantsev, Production of barium titanate and its solid solutions with lead titanate, AKIN Akad. Nauk SSSR Report, 1955.

180. M. A. Ugryumova and A. V. Sosnov, Production of barium titanate ceramic by the method of hot casting under pressure, AKIN Akad. Nauk SSSR Report, 1958.

181. A. A. Anan'eva, M. A. Ugryumova, and A. V. Sosnov, Production of piezoelectric elements by the method of hot casting under pressure, Advances in Scientific and Technical Research, 1960, No. 1, Topic 3.

182. K. I. Goncharov and V. A. Krasil'nikov, Thermal mechanical vibrations (fluctuations) in piezoelectric crystals, Izv. Akad. Nauk SSSR, Ser. Fiz. 20: 2 (1956).

183. V. M. Gotsak, A universal wide-band phase meter, Advances in Scientific and Technical Research, 1958.

184. V. V. Furdyev, Reciprocity Theorems, Gostekhizdat, 1948.

185. M. V. Kazantseva, Autocalibration of an electroacoustic transducer by a pulse method in a tube, Zh. Tekhn. Fiz. 23: 9 (1953).

186. M. V. Kazantseva, A reciprocity method for the absolute calibration of sound receivers, Dokl. Akad. Nauk SSSR 58: 8 (1947).

187. M. V. Kazantseva, Absolute calibration of electroacoustic transducers by the reciprocity method, Tr. Komis. po Akustike, Vol. 5 (1950).

188. N. S. Ageeva, Investigation of the change in the acoustic parameters of materials at ultrasonic frequencies with the aid of a pulse tube, Akust. Zh. 1: 2 (1955).

189. I. L. Kracil'shchik, Standard measurement hydrophones, Proc. of the All-Soviet Acoustic Conference, pp. 87-88, 1958.

190. A. N. Abrikosov and V. A. Kalmykov, Instruments for sound pressure measurement. Monitoring and measuring instruments for use at ultrasonic frequencies, Proc. of the Conference on Ultrasonics, pp. 129-134, 1960.

191. N. I. Shapirovskii and A. A. Gagel'gants, The possible use of ceramic sound receivers as detectors of seismic vibrations in the sea, Novosti Neft. Tekhn. Geol., No. 5, 1959.

192. E. V. Romanenko, Miniature piezoelectric sound receivers, All-Soviet Conference on the Application of Ultrasonics in Industry, 1957.

193. E. V. Romanenko, Miniature piezoelectric ultrasonic receivers, Akust. Zh. 3: 4 (1957).

194. A. A. Anan'eva, Inventor's claim No. 134294, April 16, 1960.

195. A. A. Anan'eva, A barium titanate ceramic sound receiver, AKIN Akad. Nauk SSSR Report, 1955.

196. A. A. Anan'eva, Nondirectional ceramic sound receivers, Akust. Zh. 2: 1 (1961).

197. A. A. Anan'eva, Ceramic sound receivers, All-Soviet Conference on the Application of Ultrasonics in Industry. Ultrasonic Measurements, 1957.

198. S. P. Timoshenko, Theory of Elasticity, Gostekhizdat, 1934.

199. N. N. Bezukhov, Theory of Elasticity and Plasticity, GITTL, 1953.

200. P. A. Langevin, Electroacoustic sensitivity of cylindrical ceramic tubes, J. Acoust. Soc. Am. 26 (3): 421 (1954).

201. V. A. Berezin, Double-sided sound receivers employing mechanical stress transformation, AKIN Akad. Nauk SSSR Report, 1955.

202. S. P. Timoshenko, Plates and Shells, ONTI, 1934.

203. V. A. Berezin, Sound receivers consisting of thin plates of barium titanate ceramic, AKIN Akad. Nauk SSSR Report, 1955.

204. Tomita and Yanaguti, A piezoelectric receiver, Elec. Commun. Lab. Tech. J. 7 (5): 138-145 (1959).

205. A. Lav, Mathematical Theory of Elasticity, ONTI NKT SSSR, 1935.

206. I. P. Golyamina, Thickness vibrations in polarized barium titanate plates, Akust. Zh. 1: 1 (1955).

207. E. A. G. Schow, Resonance vibrations in thick barium titanate discs, J. Acoust. Soc. Am. 28 (1): 38 (1956).

208. L. Pavlov, Multivibrations in thin plates of barium titanate ceramic, Thesis, AKIN Akad. Nauk SSSR, 1957.

209. M. Frederici, Electrostriction vibrations in barium titanate tubes, Alta Frequenza 24: 1 (1959).

210. B. D. Tartakovskii, Theory of the propagation of plane waves through a homogeneous layer, Dokl. Akad. Nauk SSSR 71 (3): 465-468 (1950).

211. A. A. Anan'eva, Investigation of barium titanate ceramic radiators, AKIN Akad. Nauk SSSR Report, 1955.

212. A. A. Anan'eva, Investigation of barium titanate ultrasonic transducer, Report of the Acoustics Laboratory, FIAN, 1951.

213. A. A. Anan'eva, 'Mosaic' type barium titanate ceramic transducer, Akust. Zh., Vol. 5, 1959.